Physical Organic Chemistry

The Fundamental Concepts

Studies in Organic Chemistry

Executive Editor

Paul G. Gassman

Department of Chemistry
The Ohio State University
Columbus, Ohio

Other Volumes in Preparation

Physical
Organic
Chemistry

The Fundamental Concepts

CALVIN D. RITCHIE
Department of Chemistry
State University of New York at Buffalo
Buffalo, New York

Marcel Dekker, Inc., New York

MARCEL DEKKER, INC.

270 Madison Avenue, New York, New York 10016

LIBRARY OF CONGRESS CATALOG CARD NUMBER: 75-24640

ISBN: 0-8247-6323-8

Current printing (last digit):
10 9 8 7 6 5 4 3 2 1

PRINTED IN THE UNITED STATES OF AMERICA

PREFACE

This book has been developed from a one-semester course offered to senior undergraduates and first-year graduates in the Department of Chemistry, State University of New York at Buffalo. It is assumed that students in such a course will have completed one and one-half years of study in organic chemistry, one year of study in physical chemistry, and will have a working knowledge of calculus and at least an acquaintance with simple differential equations. The book is not intended for "bedtime reading," nor is it intended that the instructor in the course should be able to read the text to his class rather than present lectures. I would hope that even the very good students will need pen and paper at hand in order to understand the material presented. I would be disappointed if an instructor using this text did not find ample opportunity to present his own examples to amplify and extend the principles introduced.

The purpose of the book is to introduce the most important of the fundamental concepts of physical organic chemistry. In my opinion, this requires a heavy emphasis on the physical side of the subject. I have recognized, however, that most students are more intrigued by mechanisms of reactions than by, for example, the operator formalism applied to the problem of the harmonic oscillator. For this reason, I have attempted to give the book a "story line" involving thorough investigations of mechanism. There is no intent to present a survey of mechanisms of organic reactions.

The material is roughly divided into two major mechanistic themes. Chapters 1 through 5 develop the topics of kinetics, salt, solvent, and structural effects on rates of reactions, and molecular orbital theory in the context of S_N1 and S_N2 reactions. In Chapter 5, the theme begins to swing over to acid-base reactions, and the concepts of catalysis, transition state theory, and isotope effects are then presented in this context in Chapters 6 through 9. The use of matrix algebra is introduced quite early in the context of kinetics, and the entire development of molecular orbital theory is in terms of matrices. It is hoped that these examples will be sufficient to motivate students to spend the time necessary to develop facility with this valuable tool.

My debt to my own teachers is enormous. Among these teachers, in addition to Profs. W. F. Sager and E. S. Lewis with whom I have been privileged to have had extended personal contact, I include Profs. L. P. Hammett, A. Frost, and R. G. Pearson, and S. Glasstone, whose texts

<u>Physical Organic Chemistry</u>, <u>Kinetics and Mechanism</u>, and <u>Theoretical Chemistry</u>, respectively, have greatly influenced my thinking.

Calvin D. Ritchie
Buffalo, New York

CONTENTS

Physical Organic Chemistry

The Fundamental Concepts

Chapter 1

KINETICS: INTEGRATION OF SIMPLE RATE EXPRESSIONS

1-1. INTRODUCTION

Physical organic chemistry is concerned with the study of factors which affect the rates and equilibria of organic reactions. Since reactions may involve intermediates, and since various factors, such as reaction solvent, temperature, or pressure, may affect the formation and further reaction of such intermediates in different ways, a first step in physical organic investigations is the establishment of mechanism of reaction. This step usually begins with the determination of an empirical rate law for the reaction. An empirical rate law states how the rate of formation of products depends on the concentrations of reactants, products, and catalysts of the reaction while all other variables, such as temperature, pressure, and ionic strength, are held constant.

   The use of the empirical rate law in aiding in the formulation of a reaction mechanism depends on the concept of an "elementary step" of a reaction. An elementary step of a reaction must have a rate law corresponding to the stoichiometry of that step. That is, if the reaction:

$$wA + xB \underset{k_r}{\overset{k_f}{\rightleftharpoons}} yC + zD$$

is an elementary step, then the rate law for this step must be

$$\frac{1}{z}\frac{d[D]}{dt} = \frac{1}{y}\frac{d[C]}{dt} = -\frac{1}{w}\frac{d[A]}{dt} = -\frac{1}{x}\frac{d[B]}{dt} = k_f[A]^w[B]^x - k_r[C]^y[D]^z$$

If the empirical rate law for a reaction does not correspond to the overall stoichiometry of the reaction, the reaction must involve more than one elementary step, and, therefore, must involve an intermediate.

   The determination of an empirical rate law for a reaction is not a trivial task, nor is the formulation of a mechanism consistent with a rate law

1

always easy. In this chapter, we shall examine several simple types of reactions to illustrate the principles involved in the determination and use of rate laws. The primary examples which will be examined are the nucleophilic substitution reactions at a saturated carbon:

$$Y^- + R_2 - \overset{R_1}{\underset{R_3}{C}} - X \longrightarrow R_2 - \overset{R_1}{\underset{R_3}{C}} - Y + X^-$$

These reactions are among the most thoroughly studied in physical organic chemistry, and their mechanisms have sufficient variety and subtlety that we will be able to use the same reactions as examples of the applications of other concepts to be discussed in later chapters.

Very early studies of the kinetics of nucleophilic substitution reactions of alkyl halides revealed the existence of at least two quite different mechanisms [1]. Some substrates, particularly primary and simple secondary alkyl halides, exhibited second-order kinetics and inversion of stereochemistry at the substituted carbon, while other substrates, particularly tertiary halides, exhibited first-order kinetics and predominant racemization at the substituted carbon.

1-2. SECOND-ORDER REACTIONS AND PSEUDO-FIRST-ORDER CONDITIONS

The reaction of n-butylbromide with iodide ion in acetone solution is a typical example of nucleophilic substitution at a primary carbon. The course of the reaction can be followed by observation of the appearance of products or of disappearance of reactants. It is found that the rate of the reaction obeys the rate law

$$\frac{d[n-BuI]}{dt} = -\frac{d[I^-]}{dt} = k[n-BuBr] \, [I^-] \tag{1-1}$$

This conclusion is reached by showing that the concentration of iodide ion (or of n-BuBr or n-BuI) as a function of time fits the behavior described by the integration of the rate law (1-1).

Generalizing the reaction scheme to

$$A + B \overset{k}{\longrightarrow} C + D \tag{1-2}$$

with

$$\frac{d[C]}{dt} = \frac{d[D]}{dt} = -\frac{d[A]}{dt} = -\frac{d[B]}{dt} = k[A] \, [B] \tag{1-3}$$

the integration can be accomplished by relating the concentrations from the stoichiometric expressions, i.e.,

$$[A]_0 - [A] = [B]_0 - [B] = [C] = [D] \tag{1-4}$$

where $[A]_0$ and $[B]_0$ are the concentrations of the two reactants at time equal zero. Substitution of Eq. (1-4) into Eq. (1-3) gives

$$\frac{d[C]}{dt} = k([A]_0 - [C]) ([B]_0 - [C]) \tag{1-5}$$

in which the variables may be separated to give

$$\frac{d[C]}{([A]_0 - [C]) ([B]_0 - [C])} = kdt \tag{1-6}$$

The left-hand side of Eq. (1-6) can be directly integrated for the case where $[A]_0 = [B]_0$:

$$\int \frac{d[C]}{([A]_0 - [C]) ([B]_0 - [C])} = \frac{1}{([A]_0 - [C])} + constant \tag{1-7}$$

or for the case where $[A]_0 \neq [B]_0$:

$$\int \frac{d[C]}{([A]_0 - [C]) ([B]_0 - [C])} =$$

$$\frac{1}{[A]_0 - [B]_0} \ln \left\{ \frac{[B]_0 - [C]}{[A]_0 - [C]} \right\} + const \tag{1-8}$$

The constants of integration in Eqs. (1-7) and (1-8) can be evaluated by setting $[C] = 0$ at $t = 0$, and a little algebra gives

$$\frac{1}{[A]_0 - [C]} - \frac{1}{[A]_0} = kt \qquad \text{for } [A]_0 = [B]_0 \tag{1-9}$$

and

$$\frac{1}{[A]_0 - [B]_0} \ln \left\{ \frac{([B]_0 - [C]) [A]_0}{([A]_0 - [C]) [B]_0} \right\} = kt \quad \text{for } [A]_0 \neq [B]_0 \tag{1-10}$$

The usual treatment of data involves plotting the function of $[C]$ on the left-hand side of these equations vs t to obtain a straight line plot whose slope is k. At the end of this section, some raw experimental data are presented in the problems so that practice in handling the calculations may be obtained.

In most cases where second-order kinetics are observed, the experiments are most conveniently carried out by using pseudo-first-order conditions. In the present example of the reaction of n-butylbromide with iodide ion, a large excess of iodide ion, such that $[I^-]$ remains constant within desired accuracy throughout the experiment, would be employed. Under these conditions, $[I^-]$ in the rate expression (1-1) can be treated as a constant, and using the stoichiometric relationships again, the rate expression can be directly integrated to give

$$\ln [\text{n-BuBr}] = -k[I^-]_0 t \qquad\qquad (1-11)$$

The plot of the left-hand side of Eq. (1-11) vs t gives a straight line with slope of $k[I^-]_0$, which is called the pseudo-first-order rate constant k_ψ. A series of experiments using different $[I^-]_0$, all in large excess, would serve to demonstrate the dependence of the pseudo-first-order rate constant on $[I^-]_0$.

1-3. REVERSIBLE REACTIONS

Another advantage of using a large excess of one reagent in the kinetic study of a reaction is that the reaction is sometimes forced to completion. In the discussion so far, we have considered that the reverse reaction can be neglected. In general, the reversibility of a reaction considerably complicates the kinetic expressions, although in the cases of first- and second-order reactions the integrations are still relatively straightforward. In more complex cases, it is virtually a necessity to run the reaction studies under pseudo-first-order conditions to determine the rate law.

For the simple case of a reversible first-order reaction

$$A \underset{k_r}{\overset{k_f}{\rightleftharpoons}} B \qquad\qquad (1-12)$$

the rate expression is

$$\frac{d[A]}{dt} = -k_f[A] + k_r[B] \qquad\qquad (1-13)$$

Use of the stoichiometric relationship $[A]_0 - [A] = [B]$ in Eq. (1-13) then gives

$$\frac{d[A]}{dt} = -(k_f + k_r) [A] + k_r[A]_0 \qquad\qquad (1-14)$$

which can be integrated to yield

$$\ln \left\{ \frac{k_f[A] - k_r([A]_0 - [A])}{k_f[A]_0} \right\} = -(k_f + k_r)t \tag{1-15}$$

Equation (1-15) can be written in a more convenient form by using the relationship

$$K_{eq} = k_f/k_r = \frac{[A]_0 - [A]_e}{[A]_e} \tag{1-16}$$

where $[A]_e$ is the concentration of A at equilibrium (i.e., when time equals infinity). Substitution of Eq. (1-16) into (1-15), and a little manipulation, gives

$$\ln \left\{ \frac{[A] - [A]_e}{[A]_0 - [A]_e} \right\} = -(k_f + k_r)t \tag{1-17}$$

The left-hand side of Eq. (1-17) is, in fact, in the same form that would normally be used for either reversible or nonreversible first-order reactions. Suppose, for example, that we were to follow the conversion of A into B by a spectrophotometric method, measuring absorbance I at a constant wavelength as a function of time. In general, both A and B will have some absorbance at the chosen wavelength, but will have different molar absorptivities, β_A and β_B, respectively. The observed absorbance at any time will be given by

$$I = [A]\beta_A + [B]\beta_B \tag{1-18}$$

Substituting the stoichiometric relationship into Eq. (1-18), we get

$$I = [A] (\beta_A - \beta_B) + [A]_0\beta_B, \text{ or } [A] = \frac{I - [A]_0\beta_B}{\beta_A - \beta_B} \tag{1-19}$$

Substitution of this result into either Eq. (1-11) or Eq. (1-17), with A being n-butylbromide, gives the same result:

$$\ln \left\{ \frac{I - I_e}{I_0 - I_e} \right\} = - k_\psi t \tag{1-20}$$

the only difference being that k_ψ in Eq. (1-17) is equal to the sum of the forward and reverse rate constants, while in Eq. (1-11), it is simply the forward rate constant.

Equation (1-20) is valid for the case where any measured property of the reacting system which is a linear function of concentration is substituted in place of I.

The cases of reversible second-order reactions run under both second-order and pseudo-first-order conditions are left as exercises at the end of the chapter.

1-4. MULTISTEP REACTIONS

In the above example of the reaction of n-butylbromide with iodide ion, the observed rate law is consistent with the overall stoichiometry of the reaction. The rate law, then, is consistent with a reaction mechanism consisting of a single elementary step.

The reaction of benzhydrylchloride with fluoride ion in sulfur dioxide solution [1] to produce benzhydrylfluoride and chloride ion appears, from the overall equation, to be similar to the above reaction. The kinetics, however, are quite different. In this case, the empirical rate law is first order with respect to benzhydrylchloride, but zero order with respect to the concentration of fluoride ion (i.e., it is independent of fluoride ion concentration). Since the rate law does not correspond to the overall stoichiometry, the mechanism must consist of more than one elementary step, and at least one intermediate must be involved.

In the reaction of benzhydrylchloride, simple first-order kinetics are observed even though the reaction must proceed by more than one step. There are a large number of reactions which show no integral-order rate laws at all. Many of these reactions involve the formation and decomposition of an observable intermediate, such as is formed in the course of the pyridine-catalyzed hydrolysis of acetic anhydride [2]:

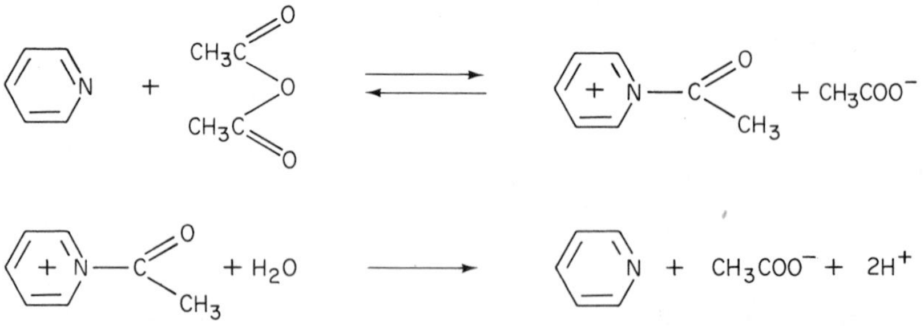

For such situations, neither the disappearance of reactants nor the appearance of products obey simple integral-order time dependences, and the concentration of intermediate first increases, then decreases as the reaction proceeds.

This general problem of simultaneous and/or consecutive reactions is, mathematically, a problem in simultaneous differential equations. Such problems are made tractable by the use of matrix algebra. Since matrix algebra will be useful to us in later sections of this book also, some of the basic principles are presented in the following section. For those already familiar with matrix manipulations, Section 1-4.1 can be skipped. For those with no prior introduction to the subject, the material in the following section should be sufficient to at least allow one to follow the further development of kinetics in Section 1-4.2.

1-4.1 Outline of Matrix Algebra

A matrix is a rectangular array of numbers, such as the following:

$$M = \begin{bmatrix} 1 & 6 \\ 7 & 9 \\ 3 & 2 \end{bmatrix} \qquad N = \begin{bmatrix} 1 & 4 & 1 \\ 2 & 1 & 1 \\ 1 & 1 & 3 \end{bmatrix}$$

$$A = \begin{bmatrix} 1 & 4 & 6 & 9 \\ 2 & 1 & 1 & 2 \\ 3 & 2 & 1 & 4 \end{bmatrix} \qquad B = \begin{bmatrix} b_{11} & b_{12} & b_{13} & b_{14} \\ b_{21} & b_{22} & b_{23} & b_{24} \\ b_{31} & b_{32} & b_{33} & b_{34} \\ b_{41} & b_{42} & b_{43} & b_{44} \end{bmatrix}$$

The elements of a matrix M are symbolized m_{ij}, where i gives the row and j gives the column in which the element is located. For the above matrices,

$$m_{12} = 6, \ m_{32} = 2, \ n_{21} = 2, \ n_{33} = 3, \ a_{34} = 4, \ \text{etc.}$$

For a square matrix, that is, one with the same number of rows and columns, the elements m_{ii} form the principal diagonal of the matrix. The dimensions of a matrix are specified as the number of rows by the number of columns. Thus, for the above matrices, the dimensions of M are 3 x 2, of N are 3 x 3, of A are 3 x 4, and of B are 4 x 4.

In the entire discussion in this section, we shall assume that all of the elements of any matrix are real numbers. Only slight modifications are necessary if one deals with complex numbers.

The basic rules of matrix algebra are defined as follows:

Equality
    M = N implies that $m_{ij} = n_{ij}$ for all i and j.

Addition

M + N = S implies that $s_{ij} = m_{ij} + n_{ij}$ for all i and j.

Subtraction

M - N = R implies that $r_{ij} = m_{ij} - n_{ij}$ for all i and j.

Multiplication

MN = P implies that $P_{ij} = \Sigma_k m_{ik} n_{kj}$ for all i and j. Note that multiplication is defined only for cases where the matrix on the left has the same number of columns as the matrix on the right has rows. It is important to note that matrix multiplication is not generally commutative: MN ≠ NM.

Some examples of these operations are given below.

$$
\begin{bmatrix} 1 & 2 & 1 \\ 2 & 1 & 2 \\ 1 & 1 & 1 \end{bmatrix}
+
\begin{bmatrix} 3 & 2 & 1 \\ 2 & 3 & 1 \\ 2 & 3 & 1 \end{bmatrix}
=
\begin{bmatrix} 4 & 4 & 2 \\ 4 & 4 & 3 \\ 3 & 4 & 2 \end{bmatrix}
$$

$$
\begin{bmatrix} 1 & 2 & 1 \\ 2 & 1 & 2 \\ 1 & 1 & 1 \end{bmatrix}
-
\begin{bmatrix} 3 & 2 & 1 \\ 1 & 2 & 3 \\ 2 & 3 & 1 \end{bmatrix}
=
\begin{bmatrix} -2 & 0 & 0 \\ 1 & -1 & -1 \\ -1 & -2 & 0 \end{bmatrix}
$$

$$
\begin{bmatrix} 2 & 1 & 1 \\ 3 & 2 & 0 \\ 1 & 2 & 1 \end{bmatrix}
\begin{bmatrix} 1 & 2 & 2 \\ 1 & 3 & 1 \\ 2 & 1 & 2 \end{bmatrix}
=
\begin{bmatrix} 5 & 8 & 7 \\ 5 & 12 & 8 \\ 5 & 9 & 6 \end{bmatrix}
$$

$$
\begin{bmatrix} 2 & 0 & 1 \\ 1 & 2 & 2 \end{bmatrix}
\begin{bmatrix} 1 & 3 \\ 2 & 2 \\ 2 & 1 \end{bmatrix}
=
\begin{bmatrix} 4 & 7 \\ 9 & 9 \end{bmatrix}
$$

$$
\begin{bmatrix} 2 & 1 & 1 \\ 3 & 2 & 0 \\ 1 & 2 & 1 \end{bmatrix}
\begin{bmatrix} 1 \\ 2 \\ 3 \end{bmatrix}
=
\begin{bmatrix} 7 \\ 7 \\ 8 \end{bmatrix}
$$

$$
\begin{bmatrix} 1 & 2 & 3 & 4 \\ 5 & 6 & 7 & 8 \\ 9 & 10 & 11 & 12 \\ 13 & 14 & 15 & 16 \end{bmatrix}
\begin{bmatrix} 1 & 0 & 0 & 0 \\ 0 & 1 & 0 & 0 \\ 0 & 0 & 1 & 0 \\ 0 & 0 & 0 & 1 \end{bmatrix}
=
\begin{bmatrix} 1 & 2 & 3 & 4 \\ 5 & 6 & 7 & 8 \\ 9 & 10 & 11 & 12 \\ 13 & 14 & 15 & 16 \end{bmatrix}
$$

The distributive and associative laws follow directly from the above definitions:

$$A(B + C) = AB + AC$$

$$(B + C)A = BA + CA$$

$$(AB)C = A(BC) = ABC$$

The number one in ordinary algebra has an analog in matrix algebra which is called an identity matrix, symbolized by I. An identity matrix is a square matrix with all ones along the principal diagonal and zeros in all other positions. The 3 x 3 identity matrix is

$$\begin{bmatrix} 1 & 0 & 0 \\ 0 & 1 & 0 \\ 0 & 0 & 1 \end{bmatrix}$$

The effect of left- or right-multiplying any matrix by the identity matrix is to leave the matrix unchanged. One of the examples of matrix multiplication given above illustrates this property. In the notation used in matrix algebra, a multiple of the identity matrix:

$$\begin{bmatrix} a & 0 & 0 & . & . & . & 0 \\ 0 & a & 0 & . & . & . & 0 \\ 0 & 0 & a & . & . & . & 0 \\ . & & & . & & & . \\ . & & & . & & & . \\ . & & & & . & . & . \\ 0 & . & . & . & . & . & a \end{bmatrix}$$

is written as the constant a. The effect of multiplying any matrix by a constant a, then, is to multiply each element of the matrix by a.

With this understanding of the meaning of a constant, you should now be able to prove that $aM = Ma$, where M is any square matrix.

In matrix algebra, as, in fact, in ordinary algebra, division is not defined except in the sense of an inverse of multiplication. Given a matrix M if one can find a matrix N such that

$$MN = I$$

then N is called a right inverse of M. For any square matrix, a left inverse is also a right inverse, and vice versa:

$$MM^{-1} = M^{-1}M = I$$

Not all matrices have inverses.

The transpose of a matrix M, written as $M^\dagger$, is the matrix whose rows are the columns of M. That is, $m^\dagger_{ij} = m_{ji}$ as, for example:

$$M = \begin{bmatrix} 1 & 2 & 3 \\ 4 & 5 & 6 \\ 7 & 8 & 9 \end{bmatrix} \qquad\qquad M^\dagger = \begin{bmatrix} 1 & 4 & 7 \\ 2 & 5 & 8 \\ 3 & 6 & 9 \end{bmatrix}$$

The following rules can be easily derived from the definition of the transpose:

$$(MN)^\dagger = N^\dagger M^\dagger, \qquad (M^\dagger)^\dagger = M, \qquad I^\dagger = I$$

An important type of matrix in many physical problems is a unitary matrix, defined as follows: If $U^\dagger U = I$, then U is a unitary matrix. If U is a square matrix, $U^\dagger U = UU^\dagger$. Notice that the definition of a unitary matrix requires that

$$\sum_k u_{ik} u^\dagger_{kj} = \sum_k u_{ik} u_{jk} = \delta_{ij}$$

where $\delta_{ij} = 1$ if $i = j$, and $= 0$ if $i \neq j$.

Another common type of matrix in physical problems is a Hermitean matrix, which has the property that the matrix is equal to its transpose. Thus, for a Hermitean matrix M, $M^\dagger = M$. Obviously a Hermitean matrix must be a square matrix.

Nearly all physical problems in which matrices are useful involve finding eigenvalues and eigenvectors of matrices. Vectors are simply matrices having a single row or a single column, and are designated as row vectors or column vectors, respectively. By this point, it should be obvious that a row vector right multiplied by a column vector is a number, and that a row vector left multiplied by a column vector is a matrix. It is common practice to designate a column vector by a small letter, such as v, and a row vector by a small letter with a dagger superscript, $v^\dagger$.

If for a matrix M a vector v can be found such that $Mv = \lambda v$, where λ is a constant, v is called an eigenvector of M and λ is an eigenvalue of M associated with the eigenvector v. For row vectors, the equations would be written

$$v^t M = v^t \lambda$$

Some examples of eigenvectors are

$$\begin{bmatrix} 2 & 1 & 1 \\ 1 & 1 & 2 \\ 1 & 2 & 1 \end{bmatrix} \begin{bmatrix} 1 \\ 1 \\ 1 \end{bmatrix} = \begin{bmatrix} 4 \\ 4 \\ 4 \end{bmatrix} = 4 \begin{bmatrix} 1 \\ 1 \\ 1 \end{bmatrix}$$

$$\begin{bmatrix} 1 & 1 & 1 \end{bmatrix} \begin{bmatrix} 2 & 1 & 1 \\ 1 & 1 & 2 \\ 1 & 2 & 1 \end{bmatrix} = 4 \begin{bmatrix} 1 & 1 & 1 \end{bmatrix}$$

$$\begin{bmatrix} 1 & 0 & 1 & 0 \\ 1 & 1 & 1 & -1 \\ 0 & 3 & -6 & 6 \\ 2 & 2 & 2 & 1 \end{bmatrix} \begin{bmatrix} 1 \\ 0 \\ 2 \\ 3 \end{bmatrix} = 3 \begin{bmatrix} 1 \\ 0 \\ 2 \\ 3 \end{bmatrix}$$

We now state without proof one of the most important theorems of matrix algebra: Any nonnull matrix has at least one nonnull eigenvector. For those interested in the proof of this theorem, Wilkenson's The Algebraic Eigenvalue Problem is an excellent reference [3].

Finding eigenvectors of a matrix is not generally an easy task, and for anything larger than a 3 x 3 matrix the computer is a virtual necessity. Nearly all modern computer centers have library programs for the eigenvector problem. In the following, we shall use a brute force method and limit our examples to 3 x 3 matrices.

The equation $Mv = \lambda v$ can be written $(M - \lambda I)v = 0$. The simultaneous equations for the elements of v implied by this equation have nonvanishing solutions for the elements of v if, and only if, the determinant $|M - \lambda I| = 0$. The expansion of the determinant gives a polynomial in λ, which must be solved for λ. The values for λ can then be substituted into the simultaneous equations to obtain the solutions for the elements of v.

Consider the matrix M whose eigenvectors we wish to find:

$$M = \begin{bmatrix} 1 & -2 & 0 \\ -1 & 2 & 0 \\ 1 & 0 & 1 \end{bmatrix} \qquad v = \begin{bmatrix} v_1 \\ v_2 \\ v_3 \end{bmatrix}$$

From this the equation $(M - \lambda I)v = 0$ can be written as follows:

$$\begin{bmatrix} 1 - \lambda & -2 & 0 \\ -1 & 2 - \lambda & 0 \\ 1 & 0 & 1 - \lambda \end{bmatrix} \begin{bmatrix} v_1 \\ v_2 \\ v_3 \end{bmatrix} = 0$$

This matrix equation requires that the three equations:

$(1 - \lambda)v_1 - 2v_2 = 0$

$-v_1 + (2 - \lambda)v_2 = 0$

$v_1 + (1 - \lambda)v_3 = 0$

be satisfied simultaneously. The equations have nonvanishing solutions for v_1, v_2, and v_3 only if

$$\begin{vmatrix} (1 - \lambda) & -2 & 0 \\ -1 & (2 - \lambda) & 0 \\ 1 & 0 & (1 - \lambda) \end{vmatrix} = 0$$

Expansion of the determinant gives the polynomial

$$(1 - \lambda)(2 - \lambda)(1 - \lambda) - 2(1 - \lambda) = \lambda^3 - 4\lambda^2 + 3\lambda = 0$$

which has solutions $\lambda = 0$, 1, or 3. The solution $\lambda = 0$ substituted into the simultaneous equations gives

$$v_1 - 2v_2 = 0, \qquad -v_1 + 2v_2 = 0, \qquad v_1 + v_3 = 0$$

which are satisfied if $2v_2 = v_1 = -v_3$. We are free to choose any value we wish for one of the vector components so long as this last relationship is used to determine the values for the other components. We shall see a little later that this is a general result, enabling us to choose the magnitudes of the components for convenience. In the present case, we arbitrarily choose $v_1 = 2$. The eigenvector associated with the eigenvalue of zero is then

$$v = \begin{bmatrix} 2 \\ 1 \\ -2 \end{bmatrix} \qquad\qquad \lambda = 0$$

The solution $\lambda = 1$ for the polynomial, substituted into the equations for the vector components gives

$$-2v_2 = 0, \qquad -v_1 + v_2 = 0, \qquad v_1 = 0$$

These equations leave us free to choose any value of v_3 that we wish, and we arbitrarily choose $v_3 = 1$ giving the eigenvector

$$v = \begin{bmatrix} 0 \\ 0 \\ 1 \end{bmatrix} \qquad\qquad \lambda = 1$$

Finally, the solution $\lambda = 3$ gives

$$-2v_1 - 2v_2 = 0, \qquad -v_1 - v_2 = 0, \qquad v_1 - 2v_3 = 0$$

Arbitrarily choosing $v_1 = 2$, we obtain the eigenvector

$$v = \begin{bmatrix} 2 \\ -2 \\ 1 \end{bmatrix} \qquad\qquad \lambda = 3$$

Quite generally, the n simultaneous equations for the vector components will consist of (n - 1) independent equations, as we have just seen for the specific case, allowing an arbitrary choice of one of the vector components. That this must be true can be seen from the following argument. Consider a matrix M with eigenvector v and associated eigenvalue λ:

$$Mv = \lambda v$$

Left-multiplying both sides of this equation by a constant a gives

$$aMv = a\lambda v$$

but we have already seen that a constant commutes with any matrix, which gives

$$Mav = \lambda av \qquad \text{or} \qquad M(av) = \lambda(av)$$

Thus, av is also an eigenvector of M with the same eigenvalue as v.

In the example just worked out, we chose the arbitrary constant a such that all elements of the vectors are small integers. Two different conventions are frequently used in physical problems. In some cases, for convenience of algebraic manipulation, the leading nonzero element of the vector is chosen as unity. This convention applied to the above example gives

$$v = \begin{bmatrix} 1 \\ 1/2 \\ -1 \end{bmatrix} \qquad v = \begin{bmatrix} 0 \\ 0 \\ 1 \end{bmatrix} \qquad v = \begin{bmatrix} 1 \\ -1 \\ 1/2 \end{bmatrix}$$

The other common convention is to "normalize the vectors to unit length."
That is, choose the arbitrary constant so that

$$\sum_i v_i^2 = 1$$

For the above example, this convention gives

$$v = \begin{bmatrix} 2/3 \\ 1/3 \\ -2/3 \end{bmatrix} \qquad v = \begin{bmatrix} 0 \\ 0 \\ 1 \end{bmatrix} \qquad v = \begin{bmatrix} 2/3 \\ -2/3 \\ 1/3 \end{bmatrix}$$

The particular matrix in the example just worked has three nonnull eigenvectors. Some matrices will have as many eigenvectors as the matrix
has columns, others will have fewer.

An extremely important type of matrix manipulation encountered in physical problems is a similarity transformation of a matrix. This is an operation on a matrix M according to the equation

$$H^{-1}MH$$

A common type of similarity transformation is a unitary transformation,
where the matrix H in the above equation is a unitary matrix, U:

$$U^{+}MU$$

Suppose that one could find a unitary matrix, each column of which is an
eigenvector of a matrix M. Then, $MU = U\Lambda$, where Λ is a diagonal matrix,
that is, one which has nonzero elements only on the principal diagonal. The
eigenvalues of M would be arranged along the principal diagonal of Λ in the
same order that the corresponding eigenvectors are arranged in U.

Consider, for example, the matrix

$$M = \begin{bmatrix} 11/6 & -4/6 & -1/6 \\ -4/6 & 14/6 & -4/6 \\ -1/6 & -4/6 & 11/6 \end{bmatrix}$$

You may verify, by carrying out the operations, that this matrix has the following eigenvectors

$$v = \begin{bmatrix} 1/\sqrt{3} \\ 1/\sqrt{3} \\ 1/\sqrt{3} \end{bmatrix} \quad v' = \begin{bmatrix} 1/\sqrt{2} \\ 0 \\ -1/\sqrt{2} \end{bmatrix} \quad v'' = \begin{bmatrix} 1/\sqrt{6} \\ -2/\sqrt{6} \\ 1/\sqrt{6} \end{bmatrix}$$

with eigenvalues $\lambda = 1$, $\lambda' = 2$, and $\lambda'' = 3$, respectively, where the eigen-vectors have been normalized to unit length. You may also verify by carrying out the operation $U^{\dagger}U$ that the matrix having these vectors as its columns is a unitary matrix.

Now, carrying out the operation, MU, we obtain

$$MU = \begin{bmatrix} 1/\sqrt{3} & 1/\sqrt{2} & 1/\sqrt{6} \\ 1/\sqrt{3} & 0 & -2/\sqrt{6} \\ 1/\sqrt{3} & -1/\sqrt{2} & 1/\sqrt{6} \end{bmatrix} \begin{bmatrix} 1 & 0 & 0 \\ 0 & 2 & 0 \\ 0 & 0 & 3 \end{bmatrix}$$

or $MU = U\Lambda$. Left-multiplying this equation by $U^{\dagger}$ we obtain a result of great usefulness in physical problems:

$$U^{\dagger}MU = \Lambda$$

As an example of the application of matrix diagonalization to the solution of simultaneous differential equations, suppose we have the equations

$$\frac{dx_1}{dt} = \dot{x}_1 = m_{11}x_1 + m_{12}x_2 + m_{13}x_3$$

$$\frac{dx_2}{dt} = \dot{x}_2 = m_{12}x_1 + m_{22}x_2 + m_{23}x_3$$

$$\frac{dx_3}{dt} = \dot{x}_3 = m_{13}x_1 + m_{23}x_2 + m_{33}x_3$$

These equations can be written in matrix form:

$$\begin{bmatrix} m_{11} & m_{12} & m_{13} \\ m_{12} & m_{22} & m_{23} \\ m_{13} & m_{23} & m_{33} \end{bmatrix} \begin{bmatrix} x_1 \\ x_2 \\ x_3 \end{bmatrix} = \begin{bmatrix} \dot{x}_1 \\ \dot{x}_2 \\ \dot{x}_3 \end{bmatrix}$$

or

$$M x = \dot{x}$$

where M is a Hermitean matrix. Since M is Hermitean, it is possible to find a unitary matrix U such that $U^\dagger MU$ is a diagonal matrix. Assume that we have, by hook or crook, found U; then the following step-by-step operations follow from our definitions and rules of matrix algebra:

$$M x = \dot{x}$$

$$MIx = MUU^\dagger x = \dot{x}$$

$$U^\dagger MUU^\dagger x = U^\dagger \dot{x}$$

Define:

$$U^\dagger MU = \Lambda, \qquad U^\dagger x = y, \qquad U^\dagger \dot{x} = \dot{y}$$

then

$$\Lambda y = \dot{y}$$

Now, since Λ is a diagonal matrix, this last equation can be written

$$\frac{dy_1}{dt} = \dot{y}_1 = \lambda_1 y_1$$

$$\frac{dy_2}{dt} = \dot{y}_2 = \lambda_2 y_2$$

$$\frac{dy_3}{dt} = \dot{y}_3 = \lambda_3 y_3$$

which can be independently integrated on inspection to give

$$y_1 = y_{10} e^{\lambda_1 t}$$

$$y_2 = y_{20} e^{\lambda_2 t}$$

$$y_3 = y_{30} e^{\lambda_3 t}$$

where y_{10}, y_{20}, and y_{30} are the constants of integration. If the equations in terms of the original variables x_1, x_2, and x_3 are desired, then operation on y with U shows the transformation

$$y = U^\dagger x$$

$$Uy = UU^\dagger x = x$$

In this example, we made use of another important theorem of matrix algebra: <u>Any Hermitean matrix can be diagonalized by a unitary transformation.</u> Thus, we know that the matrix U required in the example exists. Matrix diagonalization programs are available in the libraries of most computer centers, and even very large matrices (for example 100 x 100) can be diagonalized on modern computers in a few seconds.

Unfortunately, not all of the matrices encountered in physical problems are Hermitean. For example, returning for a moment to the kinetic problem, the reaction scheme

$$A \underset{k_{ba}}{\overset{k_{ab}}{\rightleftharpoons}} B \underset{k_{cb}}{\overset{k_{bc}}{\rightleftharpoons}} C$$

gives the matrix equation

$$\begin{bmatrix} -k_{ab} & k_{ba} & 0 \\ k_{ab} & -(k_{ba} + k_{bc}) & k_{cb} \\ 0 & k_{bc} & -k_{cb} \end{bmatrix} \begin{bmatrix} A \\ B \\ C \end{bmatrix} = \begin{bmatrix} \dot{A} \\ \dot{B} \\ \dot{C} \end{bmatrix}$$

which involves a non-Hermitean matrix except in the unlikely case that $k_{ba} = k_{ab}$ and $k_{bc} = k_{cb}$.

Another theorem of matrix algebra, however, states that <u>any matrix may be transformed to triangular form by a similarity transformation.</u> A triangular matrix is one in which all elements below the principal diagonal, or all elements above the principal diagonal, are equal to zero. We shall see that this is sufficient to allow us to solve any problem involving simultaneous and consecutive first-order reactions.

The problem of triangularizing a matrix is basically that of finding eigenvectors, and the basic theorems and methods have already been presented. The most important theorem is again an existence theorem that we have already stated: any nonnull matrix has at least one eigenvector.

Given a matrix M we find in some way an eigenvector v

$$
\begin{bmatrix}
m_{11} & m_{12} & \cdot & \cdot & \cdot & m_{1j} \\
m_{21} & m_{22} & & \cdot & \cdot & m_{2j} \\
\cdot & & \cdot & \cdot & \cdot & \cdot \\
& & & & & \\
m_{j1} & & \cdot & \cdot & \cdot & m_{jj}
\end{bmatrix}
\begin{bmatrix}
1 \\ v_2 \\ \cdot \\ \cdot \\ \cdot \\ v_j
\end{bmatrix}
= \lambda_1
\begin{bmatrix}
1 \\ v_2 \\ \cdot \\ \cdot \\ \cdot \\ v_j
\end{bmatrix}
$$

where the vector v has been normalized so that its leading nonzero element is unity. It may not always be possible to find an eigenvector of the precise form shown, since all eigenvectors may have their first elements equal to zero. For the present, we shall assume, in order to keep the development as simple as possible, that all eigenvectors used have nonzero first elements. After developing the method, we will return to this point and show that a slight modification of the method will take care of the problem.

We now construct a matrix R by replacing the first column of the identity matrix with v:

$$
R =
\begin{bmatrix}
1 & 0 & 0 & \cdot & \cdot & \cdot & 0 \\
v_2 & 1 & 0 & \cdot & \cdot & \cdot & 0 \\
v_3 & 0 & 1 & \cdot & \cdot & \cdot & 0 \\
\cdot & \cdot & \cdot & \cdot & \cdot & \cdot & \cdot \\
v_j & \cdot & \cdot & \cdot & \cdot & \cdot & 1
\end{bmatrix}
$$

You may quickly verify that the matrix R^{-1} formed by changing the signs of all elements except the first in the first column of R, i.e.,

$$
\begin{bmatrix}
1 & 0 & 0 & \cdot & \cdot & \cdot & 0 \\
-v_2 & 1 & 0 & \cdot & \cdot & \cdot & 0 \\
-v_3 & 0 & 1 & \cdot & \cdot & \cdot & 0 \\
\cdot & \cdot & \cdot & \cdot & \cdot & \cdot & \cdot \\
-v_j & \cdot & \cdot & \cdot & \cdot & \cdot & 1
\end{bmatrix}
$$

is the inverse of R; $RR^{-1} = R^{-1}R = I$.

With a little thought, you should now be able to convince yourself that the matrix N in the equation

$$MR = RN$$

is a matrix with $n_{11} = \lambda_1$, and with all other elements in the first column equal to zero. Left multiplying this equation by R^{-1}, we obtain

$$R^{-1}MR = N$$

We now consider the matrix N^r of dimension $(j - 1) \times (j - 1)$, where $j \times j$ is the dimension of N, in which the first row and column of N are omitted:

$$N^r = \begin{bmatrix} n_{22} & n_{23} & n_{24} & \cdot & \cdot & \cdot & n_{2j} \\ n_{32} & n_{33} & n_{34} & \cdot & \cdot & \cdot & n_{3j} \\ n_{42} & n_{43} & n_{44} & \cdot & \cdot & \cdot & n_{4j} \\ \cdot & \cdot & \cdot & \cdot & \cdot & \cdot & \cdot \\ n_{j2} & \cdot & \cdot & \cdot & \cdot & \cdot & n_{jj} \end{bmatrix}$$

Following exactly the same procedure as above, we find an eigenvector of N^r, v^r, with eigenvalue λ_2, construct the matrix Q^r analogous to the matrix R above, and get the matrix C^r:

$$(Q^r)^{-1}N^rQ^r = C^r$$

This matrix has λ_2 as the first element of the first column and all other elements in the first column equal to zero.

Now construct the matrix Q and the matrix Q^{-1} according to the equations

$$Q = \begin{bmatrix} 1 & 0 & 0 & \cdot & \cdot & \cdot & 0 \\ 0 & q_{11}^r & q_{12}^r & \cdot & \cdot & \cdot & q_{1i}^r \\ 0 & q_{21}^r & q_{22}^r & \cdot & \cdot & \cdot & \cdot \\ \cdot & \cdot & \cdot & \cdot & \cdot & \cdot & \cdot \\ 0 & q_{i1}^r & \cdot & \cdot & \cdot & \cdot & q_{ii}^r \end{bmatrix} = \begin{bmatrix} 1 & 0 & 0 & \cdot & \cdot & \cdot & 0 \\ 0 & & & & & & \\ 0 & & & Q^r & & & \\ \cdot & & & & & & \\ 0 & & & & & & \end{bmatrix}$$

with $i = j - 1$.

$$Q^{-1} = \begin{bmatrix} 1 & 0 & 0 & \cdot & \cdot & 0 \\ 0 & & & & & \\ 0 & & & & & \\ \cdot & & (Q^r)^{-1} & & & \\ \cdot & & & & & \\ 0 & & & & & \end{bmatrix}$$

You should now be able to satisfy yourself that the matrix $S = Q^{-1}NQ$ has only zeros below the principal diagonal in both the first and second columns.

Continuing the process by constructing the matrix $S^{r\,'}$:

$$S = \begin{bmatrix} \lambda_1 & s_{12} & s_{13} & \cdot & \cdot & \cdot & s_{1j} \\ 0 & \lambda_2 & s_{23} & \cdot & \cdot & \cdot & s_{2j} \\ 0 & 0 & & & & & \\ 0 & 0 & & & S^{r\,'} & & \\ \cdot & \cdot & & & & & \\ \cdot & \cdot & & & & & \\ \cdot & \cdot & & & & & \\ 0 & 0 & & & & & \end{bmatrix}$$

finding eigenvector $v^{r\,'}$ with eigenvalue λ_3, constructing the matrices $A^{r\,'}$ and $(A^{r\,'})^{-1}$, analogous to Q^r above, and the matrices A and A^{-1}:

$$A = \begin{bmatrix} 1 & 0 & \cdot & \cdot & \cdot & 0 \\ 0 & 1 & 0 & \cdot & \cdot & 0 \\ \cdot & \cdot & & A^{r\,'} & & \\ \cdot & \cdot & & & & \\ \cdot & \cdot & & & & \\ 0 & 0 & & & & \end{bmatrix} \qquad A^{-1} = \begin{bmatrix} 1 & 0 & \cdot & \cdot & \cdot & 0 \\ 0 & 1 & \cdot & \cdot & \cdot & \cdot \\ \cdot & \cdot & & (A^{r\,'})^{-1} & & \\ \cdot & \cdot & & & & \\ 0 & 0 & & & & \end{bmatrix}$$

we transform S to a matrix which has only zeros below the principal diagonal in the first three columns.

Stepwise continuation of the process must give after j steps a triangular matrix T which is related to the original matrix M by the equation

$$(\ldots A^{-1}Q^{-1}R^{-1})\ M\ (RQA\ldots) = T$$

This can be simplified in appearance and for manipulation:

$$H = (RQA\ldots)$$
$$H^{-1} = (\ldots A^{-1}Q^{-1}R^{-1})$$

Therefore,

$$H^{-1}MH = T$$

Having established the general method, we can now return to a consideration of the problem posed when an eigenvector has zero as its first element. In this case, one additional step must be inserted into the procedure,
in which the eigenvector is operated on with a permutation matrix. A permutation matrix is a matrix each of whose rows and columns contain one
element equal to unity and all other elements equal to zero. The identity
matrix, of course, is included in this class of matrices. Any permutation
matrix is also a unitary matrix. A permutation matrix operating on a
vector simply permutes the elements of the vector. Some examples of
permutation matrices are

$$
\begin{bmatrix} 0 & 1 & 0 & 0 & 0 \\ 1 & 0 & 0 & 0 & 0 \\ 0 & 0 & 1 & 0 & 0 \\ 0 & 0 & 0 & 1 & 0 \\ 0 & 0 & 0 & 0 & 1 \end{bmatrix}
\quad
\begin{bmatrix} 0 & 0 & 1 \\ 0 & 1 & 0 \\ 1 & 0 & 0 \end{bmatrix}
\quad
\begin{bmatrix} 0 & 0 & 1 & 0 & 0 \\ 1 & 0 & 0 & 0 & 0 \\ 0 & 0 & 0 & 1 & 0 \\ 0 & 1 & 0 & 0 & 0 \\ 0 & 0 & 0 & 0 & 1 \end{bmatrix}
$$

Suppose now that the eigenvector v of the matrix M in the example that
we just worked, had the first i elements equal to zero:

$$
v = \begin{bmatrix} 0 \\ 0 \\ \cdot \\ \cdot \\ \cdot \\ v_{i+1} \\ v_{i+2} \\ \cdot \\ \cdot \\ \cdot \\ v_j \end{bmatrix}
$$

We then construct the permutation matrix that permutes the first and
(i + 1)th elements of this vector:

$$
P = \begin{bmatrix} 0 & 0 & \cdot & \cdot & \cdot & 1 & \cdot & \cdot & \cdot & 0 \\ 0 & 1 & \cdot & \cdot & \cdot & \cdot & \cdot & \cdot & \cdot & 0 \\ \cdot & \cdot & \cdot & \cdot & \cdot & \cdot & \cdot & \cdot & \cdot & \cdot \\ 1 & 0 & \cdot & \cdot & \cdot & 0 & \cdot & \cdot & \cdot & 0 \\ \cdot & \cdot & \cdot & \cdot & \cdot & \cdot & \cdot & \cdot & \cdot & \cdot \\ 0 & \cdot & \cdot & \cdot & \cdot & \cdot & \cdot & \cdot & \cdot & 1 \end{bmatrix}
$$

The matrix P has ones along the principal diagonal at all positions except the p_{11} and $p_{(i+1)(i+1)}$ elements, which are zeros; and has zeros in all other positions except the $p_{1(i+1)}$ and $p_{(i+1)1}$ elements, which are ones.

$$Pv = v' = \begin{bmatrix} v_{i+1} \\ 0 \\ \cdot \\ \cdot \\ \cdot \\ 0 \\ v_{i+2} \\ \cdot \\ \cdot \\ \cdot \\ v_j \end{bmatrix}$$

Note: $P = P^t$ and $PP = I$. From these relationships and the original equation, $Mv = \lambda_1 v$, we operate in a stepwise fashion:

$$Mv = MPPv = \lambda_1 v$$

$$PMPPv = P\lambda_1 v = \lambda_1 Pv$$

Therefore;

$$(PMP)v' = \lambda_1 v'$$

Now, since the vector v' has a nonzero first element, we can construct the matrix R' in the above example and proceed through the steps to obtain T. The matrix H is then $(PR'QA...)$.

It should be obvious with a little more thought that the permutation step can be inserted in the process wherever necessary to allow the construction of the transformation matrices in a similar way. The above process, in fact, proves by induction the theorem that any matrix can be triangularized by a similarity transformation.

1-4.2 Kinetics of First-order Processes

We can now return to the specific problem of a reaction proceeding through a single intermediate. We will work in detail the case of a reaction which proceeds by two consecutive steps to completion, but we can start with the more general case:

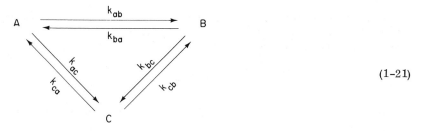

(1-21)

Defining $k_{11} = -(k_{ab} + k_{ac})$, $k_{22} = -(k_{ba} + k_{bc})$, and $k_{33} = -(k_{ca} + k_{cb})$, the kinetic expressions can be written

$$\begin{bmatrix} k_{11} & k_{ba} & k_{ca} \\ k_{ab} & k_{22} & k_{cb} \\ k_{ac} & k_{bc} & k_{33} \end{bmatrix} \begin{bmatrix} A \\ B \\ C \end{bmatrix} = \begin{bmatrix} \dot{A} \\ \dot{B} \\ \dot{C} \end{bmatrix}$$

(1-22)

or

$$Kx = \dot{x}$$

Suppose we have found the matrix H such that

$$H^{-1}KH = T$$

where T is a triangular matrix. Starting from Eq. (1-22), we can derive

$$Tx' = \dot{x}'$$

(1-23)

where

$$H^{-1}x = x', \qquad H^{-1}\dot{x} = \dot{x}'$$

(1-24)

Expansion of Eq. (1-23) gives:

$$\dot{x}_3' = \lambda_3 x_3'$$

(1-25)

$$\dot{x}_2' = \lambda_2 x_2' + t_{23}x_3'$$

(1-26)

$$\dot{x}_1' = \lambda_1 x_1' + t_{12}x_2' + t_{13}x_3'$$

(1-27)

where the λ's are the diagonal elements of the matrix T, and the t's are the off-diagonal elements.

Equation (1-25) can be integrated on inspection to give

$$x_3' = x_{30}' e^{\lambda_3 t}$$

(1-28)

which, when substituted into Eq. (1-26), gives a standard form of differential equation whose solution is [4]

$$x_2' = \left\{\frac{t_{23}x_{30}'}{(\lambda_3 - \lambda_2)}\right\} e^{\lambda_3 t} + \left\{x_{20}' - \frac{t_{23}x_{30}'}{(\lambda_3 - \lambda_2)}\right\} e^{\lambda_2 t} \qquad (1\text{-}29)$$

and now substitution of Eqs. (1-28) and (1-29) into Eq. (1-27) gives another differential equation in a standard form, whose solution is:

$$x_1' = \left\{\frac{t_{13}x_{30}'}{(\lambda_3 - \lambda_1)} + \frac{t_{12}t_{23}x_{30}'}{(\lambda_3 - \lambda_2)(\lambda_3 - \lambda_1)}\right\} e^{\lambda_3 t}$$

$$+ \left\{\frac{t_{12}x_{20}'}{(\lambda_2 - \lambda_1)} - \frac{t_{12}t_{23}x_{30}'}{(\lambda_3 - \lambda_2)(\lambda_2 - \lambda_1)}\right\} e^{\lambda_2 t}$$

$$+ \left\{x_{10}' + \frac{[t_{12}t_{23} - t_{13}(\lambda_2 - \lambda_1)]x_{30}'}{(\lambda_3 - \lambda_1)(\lambda_2 - \lambda_1)} - \frac{t_{12}x_{20}'}{(\lambda_2 - \lambda_1)}\right\} e^{\lambda_1 t} \qquad (1\text{-}30)$$

where x_{10}', x_{20}', and x_{30}' are the constants of integration.

These equations for x' can be transformed back to the original concentrations of A, B, and C in Eq. (1-21) by Eq. (1-31):

$$Hx' = x \qquad (1\text{-}31)$$

There is, however, no getting around the simple but lengthy algebra involved.

As a specific example of this method, we shall work through the case where $k_{ac} = k_{ca} = k_{cb} = 0$ in detail. This case corresponds to the following reaction scheme:

$$A \underset{k_{ba}}{\overset{k_{ab}}{\rightleftharpoons}} B \xrightarrow{k_{bc}} C \qquad (1\text{-}32)$$

which is that proposed for the S_N1 reaction. The matrix K for this problem is

$$K = \begin{bmatrix} -k_{ab} & k_{ba} & 0 \\ k_{ab} & -(k_{ba} + k_{bc}) & 0 \\ 0 & k_{bc} & 0 \end{bmatrix}$$

An eigenvector of K must satisfy the equation

$$-(k_{ab} + \lambda)v_1 + k_{ba}v_2 = 0$$

$$k_{ab}v_1 - (k_{ba} + k_{bc} + \lambda)v_2 = 0 \qquad (1\text{-}33)$$

$$k_{bc}v_2 - \lambda v_3 = 0$$

where λ can be found from expansion of the determinant:

$$\begin{vmatrix} -(k_{ab} + \lambda) & k_{ba} & 0 \\ k_{ab} & -(k_{ba} + k_{bc} + \lambda) & 0 \\ 0 & k_{bc} & -\lambda \end{vmatrix} = 0 \qquad (1\text{-}34)$$

$$(k_{ab} + \lambda)(k_{ba} + k_{bc} + \lambda)(-\lambda) + k_{ab}k_{ba}\lambda = 0$$

Equation (1-34) has as solutions for λ

$$\lambda = 0$$

and

$$\lambda = -(1/2)\left\{ (k_{ab} + k_{ba} + k_{bc}) \pm [(k_{ab} + k_{ba} + k_{bc})^2 - 4k_{ab}k_{bc}]^{1/2} \right\} \quad (1\text{-}35)$$

The solution $\lambda = 0$ substituted into Eq. (1-33) gives a null vector, so we must use one of the other solutions. Choosing λ_1 as the solution with negative sign in Eq. (1-35) gives

$$v_1 = 1$$

$$v_2 = (k_{ab} + \lambda_1)/k_{ba} \qquad (1\text{-}36)$$

$$v_3 = k_{bc}(k_{ab} + \lambda_1)/\lambda_1 k_{ba}$$

We then construct the matrix R:

$$R = \begin{bmatrix} 1 & 0 & 0 \\ v_2 & 1 & 0 \\ v_3 & 0 & 1 \end{bmatrix} \qquad R^{-1} = \begin{bmatrix} 1 & 0 & 0 \\ -v_2 & 1 & 0 \\ -v_3 & 0 & 1 \end{bmatrix}$$

With a little algebra, we find [using Eq. (1-34) to simplify terms]

$$\begin{bmatrix} \lambda_1 & k_{ba} & 0 \\ 0 & -(k_{ab} + k_{ba} + k_{bc} + \lambda_1) & 0 \\ 0 & -k_{ab}k_{bc}/\lambda_1 & 0 \end{bmatrix} = R^{-1} KR \qquad (1\text{-}37)$$

Since Eq. (1-34) shows that $-(k_{ab} + k_{ba} + k_{bc} + \lambda_1) = +k_{ab}k_{bc}/\lambda_1$, the 2 x 2 matrix formed from the last two rows and columns of the matrix in Eq. (1-37) obviously has eigenvector v^r such that

$$v^r = \begin{bmatrix} 1 \\ -1 \end{bmatrix} \tag{1-38}$$

with $\lambda_2 = +k_{ab}k_{bc}/\lambda_1$.

Constructing the matrices

$$Q = \begin{bmatrix} 1 & 0 & 0 \\ 0 & 1 & 0 \\ 0 & -1 & 1 \end{bmatrix} \qquad Q^{-1} = \begin{bmatrix} 1 & 0 & 0 \\ 0 & 1 & 0 \\ 0 & 1 & 1 \end{bmatrix}$$

and taking the transformation

$$Q^{-1}R^{-1}KRQ = T$$

$$T = \begin{bmatrix} \lambda_1 & k_{ba} & 0 \\ 0 & \lambda_2 & 0 \\ 0 & 0 & 0 \end{bmatrix} \tag{1-39}$$

and

$$RQ = H = \begin{bmatrix} 1 & 0 & 0 \\ \dfrac{k_{ab} + \lambda_1}{k_{ba}} & 1 & 0 \\ \dfrac{k_{bc} + \lambda_2}{k_{ba}} & -1 & 1 \end{bmatrix} \tag{1-40}$$

$$Q^{-1}R^{-1} = H^{-1} = \begin{bmatrix} 1 & 0 & 0 \\ \dfrac{-(k_{ab} + \lambda_1)}{k_{ba}} & 1 & 0 \\ 1 & 1 & 1 \end{bmatrix}$$

We now have the T matrix for Eq. (1-23), so Eq. (1-39) can be substituted into the general Eqs. (1-28)-(1-30) to give the integrated rate expressions for our specific case. With a great deal of simple algebra, the

assumption that the initial concentrations of B and C are zero, and application of Eq. (1-31), we obtain

$$[A] = [A]_0 \left\{ -\frac{k_{ab} + \lambda_2}{\lambda_1 - \lambda_2} e^{\lambda_1 t} + \frac{k_{ab} + \lambda_1}{\lambda_1 - \lambda_2} e^{\lambda_2 t} \right\}$$

$$[B] = [A]_0 \left\{ \frac{(k_{ab} + \lambda_2)(k_{ab} + \lambda_1)}{k_{ba}(\lambda_1 - \lambda_2)} \right\} \left\{ e^{\lambda_2 t} - e^{\lambda_1 t} \right\} \qquad (1\text{-}41)$$

$$[C] = [A]_0 - [A] - [B]$$

With these results, it is now instructive to work through several numerical examples. First, consider a case with

$$k_{ab} = k_{ba} = k_{bc} = k_0$$

The matrix K is then

$$K = k_0 \begin{bmatrix} -1 & 1 & 0 \\ 1 & -2 & 0 \\ 0 & 1 & 0 \end{bmatrix}$$

It is obvious that k_0 is simply a time scale factor that does not need to be
carried through all of the operations. We will, therefore, work with
$(1/k_0)K$. From Eqs. (1-35) and (1-38), we find that

$$\lambda_1 = (-3 - \sqrt{5})/2 = -2.62$$

$$\lambda_2 = (-3 + \sqrt{5})/2 = -0.38$$

Substitution into Eqs. (1-41) gives

$$[A] = [A]_0 \, (0.276 e^{-2.62t} + 0.724 e^{-0.38t})$$

$$[B] = [A]_0 (0.446) \, (e^{-0.38t} - e^{-2.62t})$$

These equations show that [B] will increase until the time that
$0.38 e^{-0.38t} = 2.62 e^{-2.62t}$ (by differentiation of the equations) and will then
start to decrease. Thus, in this example B reaches a maximum concentration of $0.276[A]_0$ at $t = 0.86$. At the same time, A has reached a concentration of $0.541[A]_0$. The plot of [A], [B], and [C] as functions of
time are shown in Figure 1-1a.
    Now consider the case with

$$k_{ba} = k_{bc} = 10 k_{ab} = 10 k_0$$

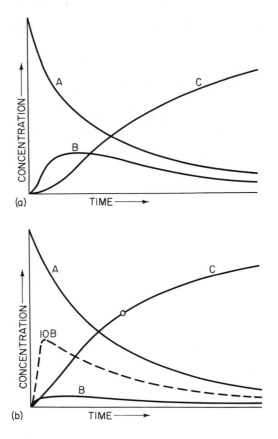

Figure 1-1. Kinetic behavior of the reaction system.

$$A \underset{k_{BA}}{\overset{k_{AB}}{\rightleftharpoons}} B \underset{k_{CB}}{\overset{k_{BC}}{\rightleftharpoons}} C$$

a) $k_{AB} = k_{BA} = k_{BC}$, $k_{CB} = 0$

b) $10k_{AB} = k_{BA} = k_{BC}$, $k_{CB} = 0$

For this example, $\lambda_1 = -20.5$ and $\lambda_2 = -0.49$. Substitutions into Eqs. (1-41) show that for this case, B reaches its maximum concentration of $0.045[A]_0$ at $t = 0.18$, at which time $[A] = 0.87[A]_0$. We may also note that the term involving $e^{\lambda_1 t}$ in the expression for $[A]$ is equal to $0.0013[A]_0$, while the term involving $e^{\lambda_2 t}$ is equal to $0.87[A]_0$ at the time when B reaches its maximum concentration. Thus, any time after ca. the first 10% reaction of A, its kinetic behavior is determined within a few percent accuracy solely by the term in $e^{\lambda_2 t}$, and, therefore, is closely following simple first-order decay. Figure 1-1b shows the behavior of the system.

A few minutes' thought should now convince you that as either k_{ba} or k_{bc} increases relative to k_{ab} (1) the time at which B reaches its maximum concentration decreases, (2) the maximum concentration of B relative to $[A]_0$ decreases, and (3) the kinetic behavior of A approaches more closely to simple first-order disappearance with apparent first-order rate constant equal to the λ of smaller magnitude.

Notice also that when the term in $e^{\lambda_2 t}$ becomes much larger than the term in $e^{\lambda_1 t}$

$$\frac{d[B]}{dt} \simeq \frac{(k_{ab} + \lambda_2)}{k_{ba}} \frac{d[A]}{dt}$$

and that $(k_{ab} + \lambda_2)/k_{ba}$ decreases as k_{ba} or k_{bc} increases relative to k_{ab}. Then, for cases where either k_{ba} or k_{bc} is much larger than k_{ab}, $d[B]/dt$ is, to a good approximation, equal to zero in comparison to $d[A]/dt$, except for an extremely short time at the beginning of the reaction.

This latter approximation is called the "steady - state approximation" or the "Bodenstein approximation." The approximation becomes more exact as $|\lambda_1|$ becomes larger than $|\lambda_2|$. As this occurs, λ_1 approaches the value $-(k_{ab} + k_{bc})$ and λ_2 approaches the value $-(k_{ab}k_{bc})/(k_{ab} + k_{bc})$.

The Bodenstein approximation is most simply made by setting the rate expression for [B] equal to zero. Thus, in the reaction scheme shown in Eq. (1-32),

$$\frac{d[B]}{dt} = 0 = k_{ab}[A] - (k_{ba} + k_{bc})\,[B]$$

or

$$[B] = \frac{k_{ab}}{(k_{ba} + k_{bc})}\,[A]$$

Since

$$\frac{d[A]}{dt} = -k_{ab}[A] + k_{ba}[B]$$

then

$$\frac{d[A]}{dt} = -\frac{k_{ab}k_{bc}}{(k_{ba} + k_{bc})}\,[A]$$

This approximation is valid for any reaction in which the concentration of the intermediate is negligably small throughout the course of the reaction; that is, if the intermediate is a highly reactive species under the conditions of the experiment.

For the generalized simple S_N1 mechanism

$$RX \underset{k_{-1}}{\overset{k_1}{\rightleftharpoons}} R^+ + X^-$$

$$R^+ + Y^- \xrightarrow{k_2} RY$$

(1-42)

application of the Bodenstein approximation gives

$$\frac{d[RX]}{dt} = -\frac{k_1 k_2 [Y^-]}{k_{-1}[X^-] + k_2[Y^-]} [RX]$$

(1-43)

Assuming that $k_2[Y^-]$ is much greater than $k_{-1}[X^-]$ throughout the reaction,

$$\frac{d[RX]}{dt} = -k_1[RX]$$

(1-44)

which is the simple first-order behavior, with rate constant independent of identity of the nucleophile Y, observed in the reaction of benzhydrylchloride with nucleophiles in sulfur dioxide solution.

In some reactions of substrates such as benzhydrylchloride, it is observed that the condition $k_2[Y^-] >> k_{-1}[X^-]$ is not satisfied. In these cases, the Bodenstein approximation still holds, but one must work with Eq. (1-43) instead of the simpler Eq. (1-44). The experimental observation which indicates this condition is a dependence of observed rate on $[X^-]$, which may either be added to the solutions initially, or may only be formed in the course of the reaction. In this latter case, a first-order plot of the rate data would show upward curvature; i.e., the apparent rate constant would decrease as the reaction proceeds. The reaction of p,p'-dimethylbenzhydrylchloride with water in 80% aqueous acetone, for example, shows this behavior.

The simple observation of a dependence of rate constant on the concentration of X^- is not sufficient, however, to verify the proposed mechanism. The rate constants for reactions generally are influenced by salt effects even in cases where the salt does not enter into the reaction scheme.

The reasons for a general dependence of rate constants on salt concentrations in solutions, and the related effects of solvents on rate constants, are analogous to those in the cases of equilibrium ratios. We shall discuss these salt and solvent effects in terms of transition state theory, which postulates a close analogy between kinetics and thermodynamics.

PROBLEMS

1. Derive the integrated rate expressions for the second-order reversible reactions

$$A + B \xrightleftharpoons[k_r]{k_f} C + D$$

and

$$A + B \xrightleftharpoons[k_r]{k_f} C$$

where $[A]_0 = [B]_0$ in both cases.

2. The reactions of aryldiazonium ions with hydroxide ion in aqueous solution involve the steps

$$ArN_2^+ + OH^- \xrightleftharpoons[k_{-1}]{k_1} syn\text{-}ArN_2OH$$

$$syn\text{-}ArN_2OH + OH^- \xrightleftharpoons[k_{-2}]{k_2} syn\text{-}ArN_2O^- + H_2O$$

Such reactions are usually studied in solutions buffered at constant pH so that pseudo-first-order kinetics are observed.
   Derive an expression that shows how the pseudo-first-order rate constant depends on hydroxide ion concentration. You may assume that the second step of the reaction is always much faster than the first step.

3. The reactions of substrates such as methylchloride in aqueous solutions actually involve reaction with both water and hydroxide ion. Describe how you would determine rate constants for each of the reactions.

4. Most techniques for the study of very fast reactions (i.e., those with half-lives of less than ca. 1 msec) involve perturbing a reaction initially at equilibrium by a rapid change in temperature or pressure. The concentrations of reactants and products must then "relax" to a new equilibrium position. Under the usual conditions, where concentrations change by a very small amount of going from the initial to the final equilibrium, first-order kinetics are always observed for one-step reactions.
   Consider the reaction

$$A + B + C + \cdots \xrightleftharpoons[k_r]{k_f} D + E + F + \cdots$$

with concentrations $[A]_0$, $[B]_0$, etc., before perturbation, and $[A]_\infty$, $[B]_\infty$, etc., after perturbation and establishment of the new equilibrium. Assuming that $|[A]_\infty - [A]_0|$, $|[B]_\infty - [B]_0|$, etc., are very small with respect to $[A]_\infty$, $[B]_\infty$, etc., show that the kinetic expressions (i.e., $d[A]/dt$) are pseudo-first-order regardless of the order of the reaction with respect to reactants or products. Describe a series of experiments that would allow you to establish the kinetic order of the reaction with respect to each species involved.

HINT: Set $a = [A] - [A]_\infty$; $da = d[A]$, etc.

What would you expect to observe if the reaction involved more than one step?

5. In Chapter 3, you will learn that the S_N1 reaction involves a more complex scheme than that shown in Eq. (1-42). In general form, the presently accepted scheme is

$$A \underset{k_{-1}}{\overset{k_1}{\rightleftharpoons}} B$$

$$B \underset{k_{-2}}{\overset{k_2}{\rightleftharpoons}} C$$

$$B \underset{k_1}{\overset{k_{-1}}{\rightleftharpoons}} A' \text{ (the enantiomer of A)}$$

$$C \underset{k_{-3}}{\overset{k_3}{\rightleftharpoons}} D + E$$

$$C \xrightarrow{k_4} \text{product}$$

$$D \xrightarrow{k_5} \text{product}$$

where B, C, and D are reactive intermediates whose concentrations do not become comparable to those of the other species.

Derive the kinetic expressions for this scheme and the integrated rate expressions for $[A]$, $[A']$, and $[product]$ under the assumption that $[E]$ is constant throughout the reaction.

6. The reaction of methyl perchlorate in aqueous solution at 0 °C was followed by placing the reaction solution in a conductivity cell and measuring the resistance of the cell as a function of time. From the data given, calculate the rate constant for the reaction from an appropriate plot.

Time (sec)	R (ohms)
0	4.0×10^5
60	39.63
100	24.01
200	12.31
500	5.296
1000	2.977
1250	2.521
1500	2.220
2000	1.853
2500	1.642
3000	1.508
15000	1.171

7. Suppose an optically active compound can racemize, rearrange with inversion, or rearrange with retention, through optically active intermediates, in the following reaction scheme:

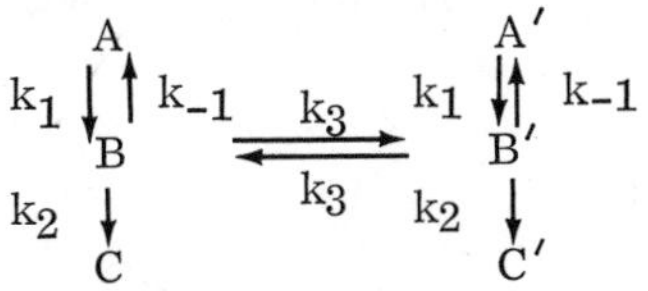

Using matrix algebra, solve the kinetic equations of the system to give integrated rate expressions for each of the species involved.

HINT: Start with the concentration vector

$$x = \begin{bmatrix} A \\ B \\ C \\ C' \\ B' \\ A' \end{bmatrix}$$

and carry out a unitary transformation of the appropriate K matrix with the matrix

$$U = \begin{bmatrix} 1/\sqrt{2} & 0 & 0 & 0 & 0 & 1/\sqrt{2} \\ 0 & 1/\sqrt{2} & 0 & 0 & 1/\sqrt{2} & 0 \\ 0 & 0 & 1/\sqrt{2} & 1/\sqrt{2} & 0 & 0 \\ 0 & 0 & -1/\sqrt{2} & 1/\sqrt{2} & 0 & 0 \\ 0 & -1/\sqrt{2} & 0 & 0 & 1/\sqrt{2} & 0 \\ -1/\sqrt{2} & 0 & 0 & 0 & 0 & 1/\sqrt{2} \end{bmatrix}$$

8. The reaction of methyl iodide with bromide ion in dimethylformamide solution at 25 °C was studied by measuring the concentration of iodide ion as a function of time. The initial concentrations of both bromide ion and methyl iodide were 3.0×10^{-3} M. From the data given below, calculate the forward and reverse rate constants for the reaction.

Time (min)	$[I^-] \times 10^3$ (M)	Time (min)	$[I^-] \times 10^3$ (M)
0	0	4.00	1.104
0.50	0.209	6.00	1.373
1.00	0.391	8.00	1.554
1.50	0.549	10.00	1.679
2.00	0.688	20.00	1.932
3.00	0.920	200.00	2.000

9. Elimination is frequently a complicating side-reaction in studies of nucleophilic substitution at saturated carbon:

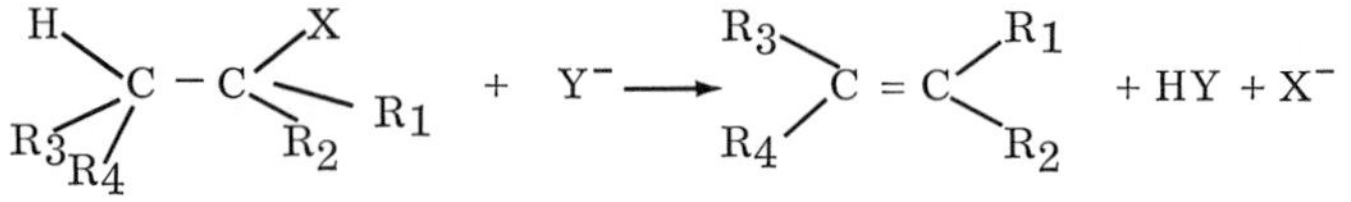

Studies of the elimination reactions of a variety of substrates show two general categories of kinetic behavior. In some cases, the rate of reaction is first-order with respect to Y^-, and in other cases the rate is independent of the concentration of Y^-. Suggest possible mechanisms for these categories.

10. In many reactions it is not possible with present techniques to measure the rate constants. As we shall see later, it is helpful to have at least relative rate constants for these cases. In some instances, the relative rate constants can be obtained by the "competition" method.

Suppose we have two substrates, S_1 and S_2, both of which react with some reagent X to give products, P_1 and P_2, respectively. For example, benzene and toluene both react with $HNO_3 + H_2SO_4$ to give nitrobenzene and nitrotoluene, respectively. The determination of relative rate constants, k_{S_1}/k_{S_2}, by the competition method involves reacting an equimolar mixture of S_1 and S_2 with the reagent and determining product ratios at some time before the reaction has appreciably changed the ratios of substrate concentrations. In some cases, this can be accomplished by isolation of products after a short enough time that only a small percentage of reaction has occurred. In other cases, an amount of reagent, X, corresponding to only a fraction of the stoichemometric amount needed to react with all of the substrates, is used.

List as many kinetic situations as you can in which the ratio of products would not give a valid measure of the relative rate constants for the substrates.

11. A frequently encountered kinetic situation is one in which an initial reactant is in rapid equilibrium with another species, which does not react directly to give product:

$$A \underset{k_{-1}}{\overset{k_1}{\rightleftharpoons}} B \qquad\qquad A \xrightarrow{k_2} product$$

with both k_1, k_{-1}, $>> k_2$. Then, throughout the reaction $[B]/[A] = K_1$, the stoichiometric relationship may be written

$$[A]_0 - (1 + K_1)\ [A] = [Product]$$

Now, consider the following two developments:

1.
$$\begin{aligned}
\frac{d[A]}{dt} &= -(k_1 + k_2)\ [A] + k_{-1}[B] \\
&= -(k_1 + k_2)\ [A] + K_1 k_{-1}[A] \\
&= -k_2[A] + (K_1 k_{-1} - k_1)\ [A]
\end{aligned}$$

but

$$K_1 = k_1/k_{-1}$$

therefore

2. $\dfrac{d[A]}{dt} = -k_2[A] \qquad$ and $\qquad [A] = [A]_0(1 - K_1)e^{-k_2 t}$ (I)

Starting with

$$[A]_0 - (1 + K_1)\ [A] = [Product]$$

$$\frac{d[Product]}{dt} = -(1 + K_1)\frac{d[A]}{dt} = k_2[A]$$

Therefore

$$\frac{d[A]}{dt} = -\frac{k_2}{(1+K_1)}\,[A]$$

$$[A] = [A]_0(1 - K_1)\,\exp[-k_2/(1+K_1)] \tag{II}$$

Equation (II) is correct. Where is the error in the derivation of Eq. (I)?

REFERENCES

1. J. Hine, Physical Organic Chemistry, McGraw-Hill Book Co., Inc., New York, New York, 1962, Chap. 6. This chapter contains an unusually complete listing of references and data concerning the kinetics of nucleophilic substitution reactions at saturated carbon.

2. A. R. Fersht and W. P. Jencks, J. Amer. Chem. Soc., 92, 5432 (1970). This paper also contains a discussion of a practical method for evaluation of the individual rate constants in the two-step reaction scheme.

3. J. H. Wilkinson, The Algebraic Eigenvalue Problem, Oxford University Press, Oxford, England, 1965.

4. M. L. Boas, Mathematical Methods in the Physical Sciences, John Wiley and Sons, Inc., New York, New York, 1966, p. 328.

Supplementary Readings

A. A. Frost and R. G. Pearson, Kinetics and Mechanism, 2nd ed., John Wiley and Sons, Inc., New York, New York, 1961, Chaps. 2, 3, and 8. These contain a thorough treatment of simple rate laws and integrations involved.

S. W. Benson, The Foundations of Chemical Kinetics, McGraw-Hill Book Co., Inc., New York, New York, 1960.

Chapter 2

KINETICS: CHARACTERIZATION OF TRANSITION STATES

2-1. THE ANALOGY BETWEEN KINETICS AND THERMODYNAMICS

There are some extremely important differences between equilibria and kinetics of reactions. For example, we have already seen that the rate expression for a reaction generally does not correspond to the overall stoichiometry of the reaction, in that some of the reactants do not appear in the rate expression. The reverse situation, called catalysis, in which some species appears in the rate expression but is not consumed or produced in the overall reaction, is also quite common.

Even in cases where stoichiometries and rate expressions do correspond, it is frequently found that the effects of structure of reactants operate in different directions on the rate and equilibrium constants for a reaction. In many cases where a reactant can produce two products by similar reactions, it is observed that the least stable product is formed faster than the more stable one. For example, in the reactions of allylic cations, produced as intermediates in the S_N1 reactions of allylic halides, the least stable of two allylically isomeric solvolysis products is often produced in the greatest amount [1]:

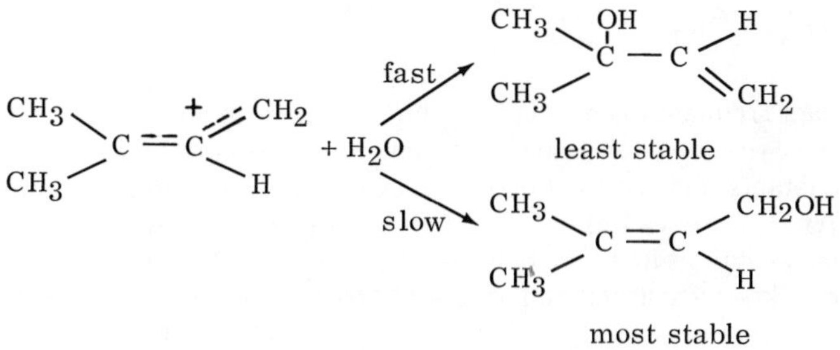

There are, however, some very striking similarities between rates and equilibria which have attracted the curiosity of chemists for many years.

The effect of temperature on reaction rate constants, for example, is empirically found to be quite well represented by the Arhennius equation:

$$k = Ae^{-E_a/RT} \tag{2-1}$$

or, in differential form

$$\frac{d(\ln k)}{d(1/T)} = - E_a/R \tag{2-2}$$

where E_a is a constant for a reaction and is called the activation energy. This equation is of the same form as the Vant-Hoff relationship for equilibrium constants:

$$\frac{d(\ln K)}{d(1/T)} = -\Delta H°/R \tag{2-3}$$

Early studies of salt effects on the rates of reactions led to the empirical relationship [2]

$$k = k° \, (\gamma_A \gamma_B/\gamma_*) \tag{2-4}$$

for reactions of ionic species A and B, in which γ_A and γ_B are the activity coefficients of A and B, respectively, and γ_* is an activity coefficient which can be calculated, using Debye-Hückel theory, for example, for a species whose charge is the algebraic sum of the charges of A and B. This equation has the appearance of an equilibrium constant expression for the following reaction:

$$A + B \rightleftharpoons A-B$$

$$\frac{(A-B)}{(A)\,(B)} = K(\gamma_A \gamma_B/\gamma_{AB}) \tag{2-5}$$

Also, although frequent exceptions are known, as pointed out above, there are many cases in which linear relationships between the logs of equilibrium constants and the logs of rate constants for a series of reactions are observed. For example, the rates of base catalyzed hydrolyses of meta- and para-substituted benzoate esters are related to the ionization constants of the like-substituted benzoic acids by the equation

$$\log k_x = \rho \, \log K_x + \text{const} \tag{2-6}$$

where k_X is the rate constant for a particular substituted benzoate ester, K_X is the ionization constant for the identically substituted benzoic acid, and ρ is the slope of the log-log plot.

Transition state theory provides a basis for understanding these similarities in rate and equilibrium constants. The central concepts of the theory are that there is some critical structure, called the transition state, which must be attained in the course of passing from reactants to products of each elementary step of a reaction, and that, once formed, this critical structure then decomposes to products with a universal rate constant of $\underline{k}T/h$, where $\underline{k}$ is Boltzmann's constant, h is Planck's constant, and T is the absolute temperature. The important hypothesis is that this critical structure, the transition state, is in thermodynamic equilibrium with reactants of the elementary step, and that we may write

$$K^* = a_*/\overset{i}{\pi} a_i = (c_*/\overset{i}{\pi} c_i) \ (\gamma_*/\overset{i}{\pi} \gamma_i) \tag{2-7}$$

where a_* is the activity of the transition state, the a_i's are activities of the reactants, the c's are the corresponding concentrations, and the γ's are activity coefficients.

With the above postulates, the rate of formation of products of the elementary step is

$$\text{rate} = d[\text{products}]/dt = (\underline{k}T/h)c_*$$
$$= (\underline{k}T/h)K^* \ (\overset{i}{\pi} c_i) \ (\overset{i}{\pi} \gamma_i/\gamma_*) \tag{2-8}$$

Defining

$$k° = (\underline{k}T/h)K^* \tag{2-9}$$

we have

$$\text{rate} = k° \ (\overset{i}{\pi} c_i) \ (\overset{i}{\pi} \gamma_i/\gamma_*) \tag{2-10}$$

which is identical to the empirical equation (2-4).

Since we have thermodynamic equilibrium between reactants and transition state, we can define free energy, enthalpy, and entropy for the conversion of reactants into transition state by the usual thermodynamic relationships involving equilibrium constants:

$$\Delta G^* = -RT \ln K^* \tag{2-11}$$

$$\Delta H^* = \frac{-Rd(\ln \ K^*)}{d(1/T)} \tag{2-12}$$

$$\Delta S^* = \frac{(\Delta H^* \ - \ \Delta G^*)}{T} \tag{2-13}$$

and, from Eq. (2-9), we can easily derive the relationships between these thermodynamic quantities and the rate constant $k°$:

$$\Delta G^* = -RT \ \ln \ k° + RT \ \ln \ (\underline{k}T/h) \tag{2-14}$$

$$\Delta H^* = -RT \ - \ \frac{R \ d(\ln \ k°)}{d(1/T)} \tag{2-15}$$

From these can be obtained the relationships between the thermodynamic quantities and the empirical constants, A and E_a, in Eq. (2-1):

$$E_a = \Delta H^* + RT \tag{2-16}$$

$$\ln A = \Delta S^*/R + \ln \ (\underline{k}T/h) + 1 \tag{2-17}$$

The primary advantage gained in the use of transition state theory is that it allows us to apply information concerning equilibria to the kinetic problem. The problem of why a reaction is fast or slow is transformed into the question of why a particular transition state is of low or high free energy, which may then be discussed in terms of enthalpy and entropy of the transition state. Similarly, the problem of salt and solvent effects on the rate constants for reactions is, by Eq. (2-10), transformed into the same problem met in equilibria, namely, that of activity coefficient behavior.

The discussion so far has been limited to consideration of the forward rate of a one-step reaction. The question of reverse rate can be handled by the fact that the transition state for the reverse reaction must be identical to that for the forward reaction. For a single elementary step:

$$A \ \underset{k_{ba}}{\overset{k_{ab}}{\rightleftharpoons}} \ B \tag{2-18}$$

we then have

$$k°_{ab} = (\underline{k}T/h)K^*_{ab} = (\underline{k}T/h)a_*/a_A \tag{2-19a}$$

$$k°_{ba} = (\underline{k}T/h)K^*_{ba} = (\underline{k}T/h)a_*/a_B \tag{2-19b}$$

and, therefore,

$$k_{ab}^{\circ}/k_{ba}^{\circ} = a_B/a_A = K_{ab} \qquad (2\text{-}20)$$

where K_{ab} is the equilibrium constant for the reaction. The same expression can be derived from the fact that the rates of the forward and reverse reactions must be equal at equilibrium.

Many reactions of interest to the organic chemist are not simple one-step reactions. In general, reactions involving simultaneous and consecutive steps must be treated in terms of the simultaneous differential equations that we have already discussed, and transition state theory must then be applied to each of the steps. In certain cases involving unstable intermediates allowing application of the Bodenstein approximation, however, the situation becomes much simpler. These are cases in which one of the steps in a series of consecutive reactions is rate determining for the overall reaction.

If, in a multistep reaction, one step is preceded only by intermediates which are in rapid equilibrium with reactants, and is followed by intermediates which go to products much faster than they revert to reactants, that step is referred to as being rate determining for the reaction. In such a case, the rate of the overall reaction is determined only by the equilibrium constant for formation of the transition state of the rate-determining step from the initial reactants. The reaction then behaves as if it proceeded by a single step with the transition state being that for the rate determining step.

Some examples may help to clarify the concept of rate-determining step. Consider first the following possible mechanism for electrophilic aromatic substitution reactions:

$$E^+ + Ar \underset{k_{-1}}{\overset{k_1}{\rightleftharpoons}} \pi\text{-complex}$$

$$\pi\text{-complex} \underset{k_{-2}}{\overset{k_2}{\rightleftharpoons}} \sigma\text{-complex} \qquad (2\text{-}21)$$

$$\sigma\text{-complex} + B \xrightarrow{k_3} ArE + BH^+$$

Treatment of both the π-complex and σ-complex by the Bodenstein approximation gives the following step-by-step development of the rate expression:

$$\frac{d[\pi]}{dt} = 0 = k_1 [Ar] [E^+] + k_{-2}[\sigma] - [k_{-1} + k_2] [\pi]$$

$$\frac{d[\sigma]}{dt} = 0 = k_2 [\pi] - (k_{-2} + k_3 [B]) [\sigma]$$

$$[\sigma] = \frac{k_2}{k_{-2} + k_3 (B)} [\pi]$$

$$[\pi] = \frac{k_1(k_{-2} + k_3 [B])}{(k_{-1} + k_2) (k_{-2} + k_3 [B]) - k_2 k_{-2}} [Ar] [E^+]$$

$$[\sigma] = \frac{k_1 k_2}{k_{-1} k_{-2} + (k_{-1} + k_2) k_3 [B]} [Ar] [E^+]$$

$$\frac{d[ArE]}{dt} = k_3 [B] [\sigma] = \frac{k_1 k_2 k_3 [B]}{k_{-1} k_{-2} + k_3 [B] (k_{-1} + k_2)} [Ar] [E^+] \qquad (2\text{-}22)$$

If $k_{-1} >> k_2$, and $k_{-2} << k_3$ (B), meaning that the π-complex is in rapid equilibrium with reactants and that the σ-complex goes to products much faster than it reverts to reactants, Eq. (2-22) simplifies to

$$\frac{d[ArE]}{dt} = \frac{k_1 k_2}{k_{-1}} [Ar] [E^+] = K_1 k_2 [Ar] [E^+] \qquad (2\text{-}23)$$

and the observed rate constant is the equilibrium constant for step 1 times the rate constant for step 2. Step 2 is then the rate-determining step of the reaction.

Step 1 of the reaction would be rate determining if $k_2 >> k_{-1}$, and k_3 [B] $>> k_{-1} k_{-2}/k_2$, which would give the rate expression

$$d[ArE]/dt = k_1 [Ar] [E^+] \qquad (2\text{-}24)$$

Changes in the concentrations of reactants can sometimes change the identity of the rate determining step and, thereby, provide evidence for the existence of more than one elementary step in a mechanism. In the present example, at least in principle, the concentration of B may be changed in such a manner that k_3 [B] $<< k_{-2}$ and k_3 [B] $<< k_{-1} k_{-2}/k_2$ at low concentrations of B, giving the rate expression

$$d[ArE]/dt = K_1 K_2 k_3 [Ar] [E^+] [B] \qquad (2\text{-}25)$$

while at higher concentrations of B, either the conditions leading to Eq. (2-24) or those leading to Eq. (2-23) would be satisfied. It is only in cases such as this, where the identity of the rate-determining step is revealed by the rate expression, that kinetic studies allow a positive identification of rate-determining step.

In particular, kinetic studies alone cannot show the existence of both the π- and σ-complex intermediates in the reaction scheme (2-21), since the rate law given by Eqs. (2-23) and (2-24) is the same. Even if it were possible, by changing temperature or activity coefficients, to change the rate-determining step from step 1 to step 2, it would be difficult to argue that the effects on observed rate constants were not simply those expected for a one-intermediate scheme.

We have already seen another example of a multistep reaction in the case of the reaction of benzhydrylchloride with fluoride ion, which proceeds by the following $S_N 1$ mechanism:

$$RX \underset{k_{-1}}{\overset{k_1}{\rightleftharpoons}} R^+ + X^-$$

$$Y^- + R^+ \xrightarrow{k_2} RY$$

This leads to the equation

$$
\begin{aligned}
k_{obs} &= \frac{k_1 k_2 (Y^-)}{(k_{-1}[X^-] + k_2[Y^-]} \\
&= k_1 \quad \text{if} \quad k_{-1}[X^-] << k_2[Y] \\
&= K_1 k_2 [Y^-]/[X^-] \quad \text{if} \quad k_{-1}[X^-] >> k_2[Y^-]
\end{aligned}
\tag{2-26}
$$

for the observed pseudo-first-order rate constant. Here again, in principle, the rate-determining step may be shifted by changing the concentration of Y^- or of X^-. Since the rate law differs for the two possible situations, kinetic evidence for the existence of the two steps can be obtained. We shall return to this case in more detail after a discussion of salt and solvent effects on rate constants, which may complicate the simple situation implied in Eq. (2-26).

It is important to realize, as already stated, that in multistep reactions in which one step is rate determining, the observed rate constant depends only on the initial reactants and the transition state for the rate-determining step. Moreover, Eqs. (2-8) through (2-20) are valid when the observed rate constant, under standard state conditions, is substituted for the k^0 in these equations. The reaction behaves as if it consisted of a single step; the activity coefficient ratio in Eq. (2-10) and the activation parameters in Eqs. (2-11)-(2-17) refer to initial reactants and transition state for the rate determining step; and the K_{ab} in Eq. (2-20) is for the overall reaction.

2-2. ACTIVITY COEFFICIENT BEHAVIOR; EFFECTS OF SALTS AND
     SOLVENTS ON THE RATES OF NUCLEOPHILIC SUBSTITUTION
     REACTIONS [2,3]

With the generalized results from the above section at our disposal, we
may now proceed to a discussion of activity coefficients, which, by Eq.
(2-10), govern the effects of salts and solvents on the rates of reactions.
    An activity coefficient is no different in principle from an equilibrium
constant, and, in fact, both are related to free energy changes by completely analogous equations.

$$\Delta G^0 = -RT \ln K \qquad\qquad\qquad\qquad\qquad\qquad (2\text{-}27)$$

Equation (2-27) relates the equilibrium constant for a chemical reaction to
the free energy change for conversion of reactants into products; while
Eq. (2-28)

$$\Delta G_t^0 = -RT \ln \gamma \qquad\qquad\qquad\qquad\qquad\qquad (2\text{-}28)$$

relates the activity coefficient of a species to the free energy change for
transferring the species from the state to which γ refers to the state in
which γ is defined as unity; that is, to the standard state. In many cases,
γ is actually determined by an equilibrium measurement such as partitioning a solute species between two immiscible solvents, or measuring the
solubility of a sparingly soluble salt in different solutions. The important
point is that activity coefficients are no less "chemical," and are no more
mysterious (nor less!) than equilibrium constants.
    It is important to grasp two facts at the outset of our discussion. First,
from Eq. (2-28) we can see that an activity coefficient of less than unity
for a species means that the free energy of transferring the species from
the state in which γ is evaluated to the standard state is positive. In more
common language, an activity coefficient of less than unity means that the
species is more "stable" than in its standard state. The opposite, of
course, is true if the activity coefficient is greater than unity. Second,
activity coefficients are not usually close to unity, as some students have
mistakenly inferred from the usual introduction to this topic through Debye-
Hückel theory. For example, the standard free energy of transfer of Li^+
from the gas phase into dilute aqueous solution is estimated as -122 kcal/
mole. Thus, at 25 °C, the activity coefficient of Li^+ in the gas phase is
10^{88} relative to a standard state in dilute aqueous solution.
    Two general rules of thumb can provide a qualitative picture of activity
coefficient behavior. Rule 1: Neutral species are more soluble in organic
solvents than in aqueous solution, while the opposite is true for ionic
species. Rule 2: Increases in salt concentration will tend to stabilize

ionic species and to destabilize neutral species in solution. There are, however, many exceptions to these generalizations associated particularly with very large ions, such as tetraalkylammonium ions, and with very polar neutral species.

In aqueous solution, the effects of salts on the activity coefficients of neutral species frequently show surprisingly good correlation by the empirical equation

$$\log \gamma_i = \lambda_{ij} c_j \tag{2-29}$$

where γ_i is the activity coefficient of the nonelectrolyte, c_j is the concentration of a particular salt, and λ_{ij} is a constant for a particular salt and nonelectrolyte. Moreover, the λ's show interesting additive properties for the cation and anion of the salt. Most salts and nonelectrolytes exhibit positive values of λ in accord with Rule 2 above, although salts with very large cations or anions are sometimes associated with negative values of λ. The λ's for polar nonelectrolytes are more negative than those for nonpolar species, and either acidic or basic nonelectrolytes, such as amines or carboxylic acids, in the presence of small cations or anions are associated with negative values of λ, probably attributable to specific Lewis acid-base interactions. The effects of salts on the activity coefficients of nonelectrolytes in nonaqueous solutions have not been examined closely.

Salt effects on the activity coefficients of ions in very dilute solutions are partially understood in terms of the Debye-Hückel theory which leads to the equation

$$\log \gamma_{\pm} = \frac{-A z_i^2 \, I^{1/2}}{1 + B I^{1/2}} \tag{2-30}$$

where A and B are constants for a given ion, solvent, and temperature, and I ($= 1/2 \, \Sigma_j c_j z_j^2$) is the ionic strength of the solution. The theoretical evaluation of A and B leave much to be desired, however, and it is perhaps more useful to view Eq. (2-30) as a special case of the empirical equation

$$\log \gamma_{\pm} = -A z_i^2 \, I^{1/2} + BI - CI^{3/2} + \cdots \tag{2-31}$$

which is an alternating power series in $I^{1/2}$, reducing to the limiting Debye-Hückel equation

$$\log \gamma_{\pm} = -A z_i^2 \, I^{1/2} \tag{2-32}$$

at very low ionic strength.

These expressions for the activity coefficients of ions fail badly at concentrations of salts where intimate ion-ion interactions, such as ion-
pairing, become important. The salt effects then become dependent on the
identity of the salt and the ion under consideration. As a very rough approximation, these special effects become important in aqueous solution
for $I > 10^{-1}$ M, and for other very polar solvents, such as methanol, dimethylformamide, or dimethylsulfoxide, at $I > 10^{-3}$ M. In even less polar
solvents, such as acetic acid or acetic anhydride, the simple behavior is
almost never observed. In these latter solvents, it is empicically found
that the rate constants for S_N1 solvolysis reactions obey the equation

$$k_{obs} = k^0 \ (1 + bc_j) \tag{2-33}$$

where c_j is the salt concentration and b is a constant dependent on the identity of the salt and the solvolyzing substrate. No theoretical justification
for this equation is apparent.

The quantitative understanding of solvent effects on activity coefficients
of solutes is even poorer than that for salt effects just discussed. The
early electrostatic theories of activity coefficient behavior focussed attention on the role of solvent dielectric constant. It is now apparent, however,
that solvent dielectric constant is of only minor, and probably indirect,
importance in determining the activity coefficients of solute ions. As examples of the types of effects observed, activity coefficients of a number of
solutes in dimethylformamide solution relative to a standard state in
methanol solution are shown in Table 2-1. The dielectric constants of
methanol and dimethylformamide are essentially equal, and it appears that
specific solvent-solute interactions are responsible for the largest changes
in activity coefficients.

The nonelectrolytes generally show small negative free energies of
transfer from methanol to DMF which may be partially attributed to the
lower "internal pressure" of DMF. Internal pressure is a measure of the
ease of cavity formation in a solvent for reception of a solute particle, and
is a function of solvent-solvent interactions. Strongly associated liquids,
particularly hydrogen-bonded liquids such as water and alcohols, have high
internal pressures which work against the introduction of solutes. Hildebrand has shown that internal pressure may be approximated as the heat of
vaporization per ml of liquid [4].

Cations are slightly more stabilized in DMF than are neutral molecules,
and the smaller cations are more stabilized than larger ones. The effect
can rationally be attributed to interaction of the cations with either the
oxygen or nitrogen lone pairs of dimethylformamide giving a slightly more
favorable interaction than that of the cations with the lone pair in methanol.

TABLE 2-1

Activity Coefficients in Dimethylformamide Solution
Relative to Standard State in Methanol Solution [3]

Solute	Log γ	Solute	Log γ
CH_3COO^-	11.9	K^+	-1.4
Cl^-	9.2	Cs^+	-0.9
Br^-	7.6	$(C_6H_5)_4As^+$	(0)
N_3^-	7.6	CH_3Cl	-0.4
$p-CH_3C_6H_4SO_3^-$	5.9	CH_3Br	-0.3
I^-	5.3	CH_3I	-0.5
SCN^-	5.4	t-Butylchloride	-0.2
Picrate$^-$	2.0	Benzylchloride	0.1
ClO_4^-	2.0	$p-O_2NC_6H_4I$	-1.2
$(C_6H_5)_4B^-$	(0)	$(C_6H_5)_4C$	-1.6
Na^+	-2.2		

Anions show the greatest free energies of transfer, and also show the greatest variation in this quantity, with small anions showing much greater effects than large anions. There is little doubt that the major factor causing the observed effect is hydrogen bonding of methanol to the anion, providing much stabilization to the ion in methanol that is not available in dimethylformamide solution.

In general summary, it appears that the major factors which influence solvent effects on activity coefficients are hydrogen bonding of solvent to solute, Lewis acid-base interactions between solvent and solute, electrostatic solvent-solute interactions involving ion-dipole or dipole-dipole forces, and solvent-solvent interactions reflected in the internal pressure of the solvent.

The principles discussed here combined with Eq. (2-10) allow one qualitatively to understand the effects of salts and solvents on the rates of various reactions where the transition states have been relatively well characterized. A more common application of the principles, however, is in the opposite sense; that of characterizing transition states from studies of salt and solvent effects on reaction rates.

Thus, for example, the fact that the rate of an S_N2 reaction of azide ion with methyl iodide is increased by a factor of $10^{4.1}$ on changing solvent from methanol to dimethylformamide can be used to show, in conjunction with Eq. (2-10) and the data in Table 2-1, that the log γ_* for the transition state is 3.0. Since this number is much smaller than that for iodide ion, the leaving iodide ion at the transition state has not attained a charge distribution similar to that of free iodide ion. Comparison with other data in the table indicates that the transition state is behaving like a very large delocalized anion. Similar types of studies for S_N1 reactions indicate that the transition states for these reactions are very similar to ion pairs such as tetraalkyl ammonium halides.

Unfortunately, very little quantitative data of the type shown in Table 2-1 is available, and most discussions of salt and solvent effects can only make qualitative use of the principles discussed above. Other empirical approaches to the problem are, however, available.

For S_N1 reactions, two model processes have been used to arrive at a quantity called solvent ionizing power, which is a measure of the activity coefficient ratio in Eq. (2-10).

Solvent Y values are determined by measurement of the rate of solvolysis of t-butylchloride in various solvents, and are defined as follows [5]:

$$Y = \log(k_S/k_0)_{t\text{-BuCl}} \qquad\qquad (2\text{-}34)$$

where k_S is the rate constant for solvolysis of t-butylchloride in the solvent under consideration, and k_0 is that for the same reaction in the standard solvent, 80:20 (v:v) ethanol:water. Comparison of Eq. (2-34) with Eq. (2-10) shows that the Y values are equal to log $(\gamma_{t\text{-BuCl}}/\gamma_*)$ with the standard state defined in 80:20 ethanol:water solvent. The Y values for some common solvents are listed in Table 2-2.

If the major feature of the change from reactant to transition state in S_N1 reactions is simply separation of charge, and there is little specific interaction of solvent with the leaving group or with the developing carbonium ion, we might expect that all S_N1 reactions would show similar solvent effects on rates, and that an equation such as

$$\log (k/k_0) = mY \qquad\qquad (2\text{-}35)$$

where m would be a function of the extent of charge development at the transition state for the particular substrate whose rate is measured, would correlate a wide variety of these reactions. Equation (2-35) does indeed correlate many S_N1 reactions, but significant deviations are found when a wide range of solvents are considered. The equation is most precise when a range of mixtures of a two-component solvent, such as ethanol-water mixtures, are considered. Significantly, the m values for all substrates

TABLE 2-2

Solvent Parameters for Nucleophilic Substitution Reactions [5,6,8]

Solvent	N	Y	Z
2-Propanol	0.09	-2.73	76.3
Ethanol	0.09	-2.03	79.6
Methanol	0.01	-1.09	83.6
80% Ethanol[a]	0.00	0.00	84.8
56% Acetone[a]	-0.47	0.9	86.5
50% Ethanol[a]	-0.20	1.6	-
Water	-0.26	3.49	94.6
50% Dioxane[a]	-0.41	1.36	-
CH_3COOH	-2.05	-1.65	79.2
HCOOH	-2.05	2.05	-
CF_3COOH	-5.55	4.5	-
DMF[b]	-	-	68.5
DMSO[c]	-	-	71.1
HMPA[d]	-	-	62.8
Acetonitrile	-	-	71.3
Sulfolane[e]	-	-	77.5

[a] The other component of the solvent is water.
[b] DMF = dimethylformamide.
[c] DMSO = dimethylsulfoxide.
[d] HMPA = hexamethylphosphoramide.
[e] Sulfolane = tetrahydrothiophene-1,1-dioxide.

which on other evidence are believed to react by the S_N1 mechanism are very close to unity, varying only from ca. 0.8-1.2.

Another measure of solvent ionizing power, Z values, is based on the wavelength of light absorbed by a charge-transfer complex of N-methyl-4-carbomethoxypyridinium iodide in various solvents [6]. The absorption of light by a charge-transfer complex is associated with an electronic transition in which negative charge is partially transferred from the iodide ion to

the pyridinium cation in an ion pair. The transition involves the transformation of an initial state consisting of a negative ion in close contact
with a positive ion to an excited state in which the two ions have partially
neutralized each other. The energy required for the transition, which is
related to the wavelength of light absorbed by $E = h\nu$, depends on the solvent stabilization of the initial state and the final state. Since the process
is similar to the reverse of the transformation of a polar C—X bond into a
charge-separated bond in the conversion of reactant to transition state in
the S_N1 reaction, solvent effects on the two processes are expected to be
inversely related to one another. That is, as the energy required for the
charge-transfer excitation increases, the energy required for the transformation of reactant to transition state for an S_N1 reaction should decrease.
The Z values are defined as equal to $h\nu$, expressed in units of kcal/mole,
where ν is the frequency of light absorbed by the charge transfer complex
in the particular solvent. Table 2-2 contains a list of these values.

From the above discussion, we might expect a linear relationship between Y and Z values, and a rough correlation is found. As for the case of
correlation of solvolysis rates by Eq. (2-35), the relationship between Y
and Z is most precise when a range of mixtures of a two-component solvent
system is considered, and is least precise, failing rather badly, when a
range of pure solvents is considered.

The S_N1 solvolysis reactions do not involve solvent as a formal reactant
in the rate-determining step, and it is for this reason that we have been
able to discuss solvent effects on these reactions in terms of activity coefficients only. The S_N2 solvolysis reactions, however, involve solvent as
a formal reactant in the rate-determining step. Solvent effects on rate constants for these reactions, which are determined as pseudo-first-order rate
constants because of excess of solvent, will reflect the reactivity of the
solvent as well as solvent effects on the activity coefficients of reactant and
transition state. It is by no means certain that the two effects can be separated, although some partially successful attempts to correlate solvent
effects on S_N2 reactions assign values to the two different functions of
solvents. These attempts have all taken the form of a four parameter
equation such as [7]

$$\log (k/k_0) = sN + mY \qquad\qquad (2\text{-}36)$$

where N is a measure of solvent nucleophilicity, s is a measure of the
susceptibility of the particular substrate to solvent nucleophilicity, and m
and Y are similar to the parameters in Eq. (2-35). One of several alternative approaches to the evaluation of the parameters in Eq. (2-36) uses
methyl tosylate as the model S_N2 substrate in much the same way that t-
butyl chloride was used in Eq. (2-34). An m value of 0.30 and an s value
of 1.00 are assigned to methyl tosylate, and Y values from Eq. (2-34) allow

N to be evaluated from the measured rates of solvolysis of methyl tosylate
in various solvents [8]. Some of the values obtained by this procedure are
shown in Table 2-2.

2-3. THE MASS LAW EFFECT IN S_N1 REACTIONS

In the earlier parts of this chapter, we discussed the fact that the two-step
S_N1 mechanism should show a change in rate-determining step for the reaction of an RX molecule as the concentration of X^- is varied. For a solvolysis reaction, the application of the Bodenstein approximation to R^+
gives the rate expression

$$k_{obs} = \frac{k_1 k_2}{k_{-1}[X^-] + k_2} \qquad (2-37)$$

where k_2 is the first-order rate constant for reaction of R^+ with solvent.
Defining $\alpha = k_{-1}/k_2$, Eq. (2-37) can be rearranged to give

$$k_{obs} = \frac{k_1}{\alpha[X^-] + 1} \qquad (2-38)$$

The observation of rate dependence on the concentration of X^- as given
by Eq. (2-38) would not only furnish evidence for the two-step nature of the
reaction, but would demonstrate that the intermediate formed in the reaction
can react with X^- to reform reactant, thereby strongly supporting the postulate of a "free" carbonium ion as the intermediate. As we have discussed
in the immediately preceding section, however, ionic strength alone is expected to have some effect on the observed rate constant, and we might
expect some difficulty is disentangling this general salt effect from the mass
law effect of Eq. (2-38).

A typical study of the mass law effect was carried out for the solvolysis
of diphenyldichloromethane in 75% aqueous acetone [9]. General salt effects were minimized by maintaining the ionic strength of all solutions at
5×10^{-2} M by addition of appropriate amounts of KBr along with the common
ion salt, KCl. Note that the use of KBr should cause no complications even
if the carbonium ion reacted with Br^- since RBr is expected to solvolyze
faster than the original substrate. The observed rate constant was found to
vary with $[Cl^-]$ in accord with Eq. (2-38), with $\alpha = 37$ M^{-1} and $k_1 = 4.82 \times
10^{-4}$ sec^{-1} at 0 °C.

Actual observations of kinetic behavior in accord with Eq. (2-38) have
only been made in cases where the carbonium ion intermediate is expected
to be relatively stable, such as in the reactions of benzhydryl halides. No
mass law effect has been observed for the solvolysis of simple tertiary

systems such as t-butylhalides, presumably because the value of α is so small in these cases that halide concentrations high enough to make $\alpha[X^-]$ approach a value of unity cannot be reasonably attained. Some typical values of α for benzhydryl systems in 80% aqueous acetone are [10]: benzhydrylchloride, $\alpha = 10$; p-methylbenzhydrylchloride, $\alpha = 35$; p,p'-dimethylbenzhydrylchloride, $\alpha = 69$.

Some caution in the interpretation of these experiments is necessary, however. In addition to the complications possible from specific salt effects and ion pairing of the common ion at the high salt concentrations used, there are further possible difficulties signalled by an unusual observation on the solvolysis of 1-anisyl-2-propyltosylate in acetic acid solution [11]. Addition of lithium tosylate to the solvolysis solution in this case produced only a normal salt effect in accord with Eq. (2-33). If, however, lithium tosylate was added to a solution already containing lithium perchlorate, a mass law effect similar to that of Eq. (2-38) was observed. This observation requires a drastic revision in the mechanism of the S_N1 reaction from that which we have written so far, and we shall return to the problem in the next chapter.

2-4. STEREOCHEMISTRY AND ISOTOPE EXCHANGE IN NUCLEOPHILIC SUBSTITUTION REACTIONS

In characterizing the mechanism of a reaction, we would like to be able to go beyond the simple detection of elementary steps and actually determine the structure of the transition state for the reaction. A major step in this direction can sometimes be made by combining stereochemical studies with the more usual kinetic studies already discussed.

At the beginning of our discussion of nucleophilic substitution at saturated carbon, we mentioned that the two classes of reaction, the S_N1 and S_N2 categories, showed different stereochemical behavior. The demonstration of the stereochemical courses of these reactions involved the use of some rather clever but simple techniques which have wide utility in the study of reaction mechanism. The preparation of pure optical enantiomers is a difficult, time-consuming task which these techniques avoid while still allowing conclusions about the absolute stereochemistry of the reactions.

The general method involves the simultaneous measurement of rates of loss of optical activity and of isotope incorporation into reactant. As an example of the method, consider first the reaction of a substrate RI with iodide ion trace labelled with radioactive iodide ion:

$$RI + {}^*I^- \xrightleftharpoons{k} R{}^*I + I^-$$

Note that the reaction is reversible, and that, if the extremely small isotope effect on rate constant is neglected, the forward and reverse rate constants are equal regardless of mechanism. Since tracer levels of the radioactive iodide are used, the concentration of radioactive iodide ion is very small in comparison to the total concentration of iodide ion.

Let us define two quantities, x and y, as follows:

$$x = \frac{[*I^-]}{[I^-] + [*I^-]} = [*I^-]/[I^-] \tag{2-39}$$

$$y = \frac{[R*I]}{[RI] + [R*I]} = [R*I]/[RI] \tag{2-40}$$

Under the conditions where iodide ion is in large excess over RI, the quantity y will have a value of zero at time equals zero and will approach the value of x as the reaction approaches equilibrium. Measurement of the progress of the reaction is then carried out by measurement of y as a function of time. The step-by-step derivation of the integrated rate expression is as follows:

$$dy/dt = \frac{1}{(RI)} \frac{d[R*I]}{dt}$$

$$\frac{d[R*I]}{dt} = k[*I^-] [RI] - k[I^-] [R*I]$$

$$= k[I^-] (x[RI] - [R*I])$$

$$= k[I^-] [RI] (x - y)$$

$$dy/dt = k[I^-] (x - y)$$

$$dy/(x-y) = k[I^-] dt$$

$$\ln (x - y) - \ln (x) = -k[I^-]t = -k_{ex}t \tag{2-41}$$

where k_{ex} is the observed pseudo-first-order rate constant for exchange.

Now consider the reaction of an optically active compound RI with iodide ion:

$$d-RI + I^- \xrightarrow{\ \ k'\ \ } 1-RI + I^-$$

Again the forward and reverse rate constants must be equal. The iodide ion may attack RI, with iodide exchange, to produce either d-RI or 1-RI. By following optical activity of the solution, however, we will observe only those reactions which invert the configuration at carbon. The reaction solution is placed in a polarimeter, and we observe the rotation of light α as a function of time. The rotation of light by the solution is given by the equation

$$\alpha = \alpha_0([d-RI] - [1-RI]) \tag{2-42}$$

where α_0 is the molar rotation of light by the optically pure d-RI. We shall
see that we do not need to know α_0 in this experiment. The derivation of
the integrated rate equation is

$$\frac{d\alpha}{dt} = \alpha_0 \frac{d([d\text{-RI}] - [l\text{-RI}])}{dt} = \frac{2\alpha_0[d\text{-RI}]}{dt}$$

$$= -2\alpha_0 k'[I^-]\,([d\text{-RI}] - [l\text{-RI}])$$

$$= -2\alpha_0 k'[I^-]\,(\alpha/\alpha_0) = -2k'[I^-]\alpha$$

$$\ln(\alpha/\alpha_{t=0}) = -2k'[I^-]t = -k_{rac}t \tag{2-43}$$

where k_{rac} is the observed pseudo-first-order rate constant for loss of
optical activity.

If we carry out the measurement of both radioactive iodide incorporation
and the loss of optical activity of an optically active compound RI, we then
obtain both k_{ex} and k_{rac} from Eqs. (2-41) and (2-43). If every iodide exchange converts d-RI into l-RI, or l-RI into d-RI, then k in Eq. (2-41) is
identical to k' in Eq. (2-43), and k_{rac} must equal $2k_{ex}$. Reversing the
argument, if $k_{rac} = 2k_{ex}$, then every exchange of iodide ion must have produced inversion of configuration. A little thought produces the following
list of possibilities:

1. If $k_{rac} = 2k_{ex}$, exchange gives 100% inversion.
2. If $k_{rac} = k_{ex}$, exchange gives 100% racemization.
3. If $2k_{ex} > k_{rac} > k_{ex}$, exchange gives dominant inversion.
4. If $k_{rac} < k_{ex}$, exchange gives dominant retention.

Some typical experimental results for S_N2 reactions are shown in Table
2-3.

TABLE 2-3

Rate Constants for Racemization and Exchange [12]

Reaction	k_{ex}[a]	k_{rac}[a]
1-Phenylethyl bromide + *Br$^-$ in acetone	8.7×10^{-4}	15.9×10^{-4}
2-Octyliodide + *I$^-$ in acetone	13.6×10^{-4}	26.2×10^{-4}
α-Bromopropionic acid + *Br$^-$ in acetone	5.2×10^{-4}	10.5×10^{-4}

[a] Values reported are the pseudo-first-order rate constants divided by
the anion concentration. Units are $M^{-1}\,sec^{-1}$.

For each of the cases shown in Table 2-3, $k_{rac} = 2k_{ex}$ within the experimental accuracy of the determination. The same result has been obtained in every case where clean second-order kinetics of the nucleophilic substitution reaction is observed. The clean inversion of configuration in S_N2 reactions is observed even in cases that are loaded to encourage front-side attack of the incoming nucleophile. For example, the reactions

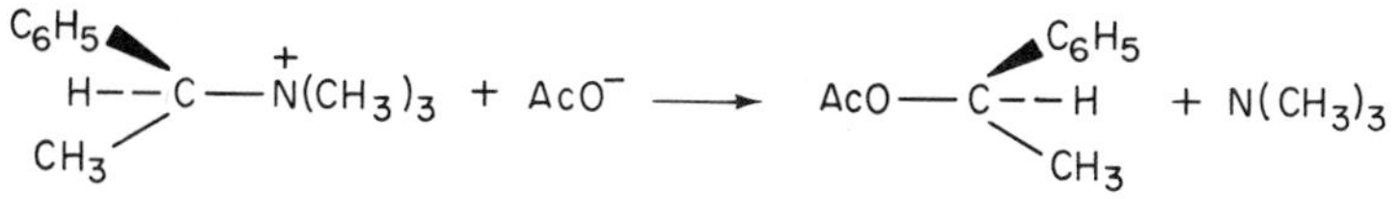

$$C_6H_5\text{---}\overset{H}{\underset{CH_3}{\text{C}}}\text{---}\overset{+}{N}(CH_3)_3 + AcO^- \longrightarrow AcO\text{---}\overset{C_6H_5}{\underset{CH_3}{\text{C}}}\text{---}H + N(CH_3)_3$$

$$C_6H_5\text{---}\overset{H}{\underset{CH_3}{\text{C}}}\text{---}\overset{+}{S}(CH_3)_2 + Br^- \longrightarrow Br\text{---}\overset{C_6H_5}{\underset{CH_3}{\text{C}}}\text{---}H + S(CH_3)_2$$

give 100% inversion of configuration [13], even though the positive charges on nitrogen or sulfur should strongly favor approach of the negatively charged nucleophile from the front side.

The inversion of configuration is one of the most clearly established features of the S_N2 reaction, and it leads to the following picture of the transition state for these reactions:

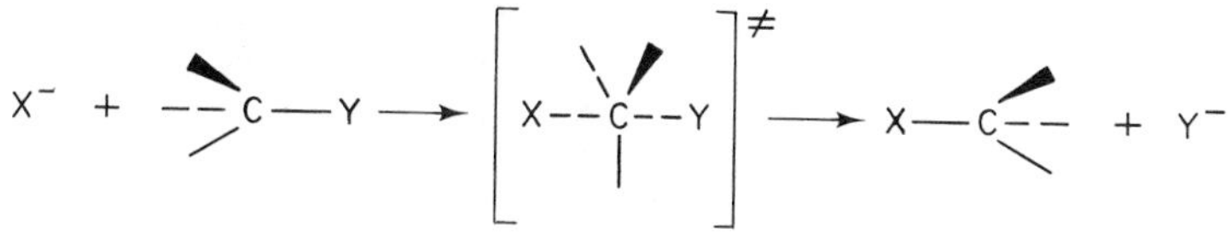

$$X^- + \text{---}\overset{}{\text{C}}\text{---}Y \longrightarrow \left[X\text{--}\overset{}{\text{C}}\text{--}Y \right]^{\neq} \longrightarrow X\text{---}\overset{}{\text{C}}\text{---} + Y^-$$

The same picture has also been found in recent quantum-mechanical calculations of potential energy surfaces [14].

The stereochemistry of S_N1 reactions is not as simple as that for the S_N2 reactions. The only clear point to emerge from many studies of the stereochemistry of S_N1 reactions is that these reactions usually involve substantial racemization. In some cases, however, retention of configuration is dominant and in others, inversion is dominant. Some typical examples are the following:

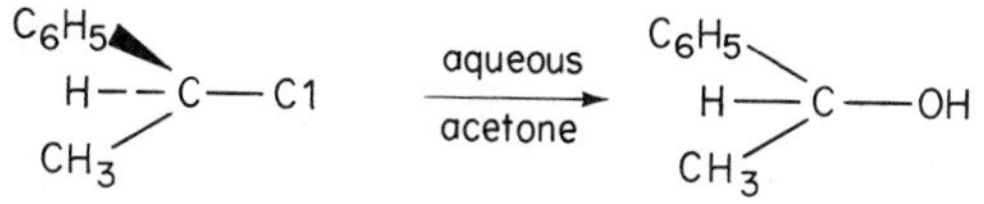

$$C_6H_5\text{---}\overset{H}{\underset{CH_3}{\text{C}}}\text{---}Cl \xrightarrow[\text{acetone}]{\text{aqueous}} C_6H_5\text{---}\overset{H}{\underset{CH_3}{\text{C}}}\text{---}OH$$

97% racemic

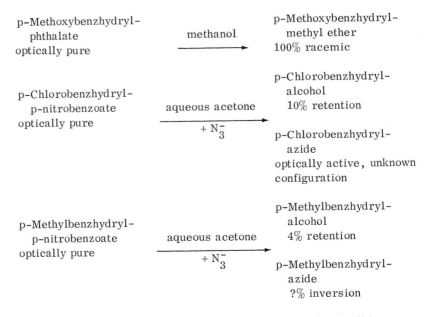

There is good reason to believe that a true carbonium ion will have a planar arrangement of the three groups bound to it, and, therefore, that products formed from a carbonium ion intermediate in S_N1 reactions would be completely racemized. The observations of incomplete racemization, alone, furnish evidence that the simple mechanism of S_N1 reactions which we have considered thus far cannot be completely correct. These observations along with some of the observations on salt effects, which we have already mentioned, require a reexamination of the entire mechanistic scheme for the S_N1 reaction.

PROBLEMS

1. The first-order rate constants for solvolysis of a compound RX in aqueous solution were found to vary with temperature as follows:

k_{obs} (sec^{-1})	T (°C)
1.20×10^{-4}	5.0
1.64×10^{-4}	10.0
2.72×10^{-4}	20.0
4.70×10^{-4}	30.0
7.90×10^{-4}	40.0
1.23×10^{-3}	50.0

Calculate E_a, ln A, ΔG^*, ΔH^*, and ΔS^* for the reaction at 25 °C.

2. Transition states are always less stable than the reactants or products of a reaction. That is, ΔG^* and ΔH^* are always positive for an elementary step of a reaction.

   Suppose you were to observe that the second-order rate constant for the reaction A + B $\rightleftharpoons$ C decreased as temperature was increased. Propose an explanation of the observation.

3. Aniline and 2,4-dinitrochlorobenzene in ethanol solution react to form 2,4-dinitrodiphenylamine and hydrochloric acid. The reaction is followed under pseudo-first-order conditions with aniline in excess, by observing the rate of production of chloride ion.

   A series of experiments at 25 °C, all with the initial concentration of 2,4-dinitrochlorobenzene at 2.00×10^{-2} M, gave the following data:

[Aniline] (M)	k_ψ (sec^{-1})
2.00×10^{-1}	1.47×10^{-5}
3.00×10^{-1}	2.12×10^{-5}
4.00×10^{-1}	2.72×10^{-5}
6.00×10^{-1}	3.76×10^{-5}
8.00×10^{-1}	4.65×10^{-5}
1.00	5.44×10^{-5}
1.50	6.98×10^{-5}
2.00	8.15×10^{-5}

   Suggest a mechanism for the reaction and evaluate as many of the rate and/or equilibrium constants as possible.

4. The reaction of p-tolyldiazonium ion with 2-napthylamine-6-sulfonate in aqueous solution at 10 °C, with added potassium chloride to maintain ionic strength, gave the following second-order rate constants as functions of ionic strength.

I (M)	k (M^{-1} sec^{-1})
0.05	0.189
0.10	0.165
0.15	0.153
0.20	0.144
0.30	0.135

0.40	0.131
0.50	0.130

Calculate the rate constant which would be obtained at $I = 0$ by the use of Eq. (2-31) with $A = 1.016$.

5. The hydrolysis of benzhydrylchloride in 80% aqueous acetone at 0°C was studied in the presence of varying concentrations of potassium chloride. The ionic strength of the solutions was maintained at 0.1 M by addition of the appropriate amounts of potassium bromide. The following data were obtained:

[KCl] (M)	k_{ψ} (sec^{-1})
0.01	2.56×10^{-6}
0.02	2.35×10^{-6}
0.04	2.01×10^{-6}
0.07	1.66×10^{-6}
0.085	1.53×10^{-6}

Calculate k_1 and α of Eq. (2-38).

6. The reaction of ethylene oxide in water is believed to follow the following mechanism:

$$CH_2\overset{O}{-}CH_2 + H^+ \underset{}{\overset{fast}{\rightleftharpoons}} CH_2\overset{\overset{H}{\overset{|}{O^+}}}{-}CH_2$$

$$CH_2\overset{\overset{H}{\overset{|}{O^+}}}{-}CH_2 + OH^- \xrightarrow[S_N2]{slow} HOCH_2CH_2OH$$

If the reaction is studied in buffered aqueous solution, how do you expect the observed pseudo-first-order rate constants to vary with pH? What salt effects would you expect to observe on the rate of reaction?

7. The first-order rate constants for solvolysis of a tertiary alkylbromide, RBr, were measured in several solvents:

Solvent	k_ψ (sec^{-1})
50% Ethanol	6.0×10^{-4}
80% Ethanol	1.8×10^{-5}
Ethanol	2.1×10^{-7}
50% Dioxane	3.6×10^{-4}
CF$_3$COOH	3.5×10^{-1}

Estimate the rate constants for solvolysis of RBr in isopropanol, acetic acid, and formic acid.

8. The rate constant for the S_N2 reaction of Br$^-$ with CH$_3$OTs in methanol at 25°C is 2.5×10^{-5} M^{-1} sec^{-1}, and in dimethylformamide is 3.1×10^{-2} M^{-1} sec^{-1}. Assuming that the solubility of CH$_3$OTs is the same in methanol and in dimethylformamide, evaluate γ_* for the transition state of the reaction in dimethylformamide, where the standard state is in methanol solution.

9. The rate constant for reaction of p-nitrophenyliodide with azide ion in methanol is 3.2×10^{-10} M^{-1} sec^{-1}, and in dimethylformamide is 5.0×10^{-6} M^{-1} sec^{-1}, both at 25°C. Calculate the free energy of transfer of the transition state of the reaction from methanol to dimethylformamide solution.

10. The reaction of an optically pure α-bromocarboxylic acid in methanol solution produces the corresponding α-methoxycarboxylic acid. The α-bromocarboxylic acid has an ionization constant of 1.0×10^{-9} M in methanol solution. The solvent ionization constant of methanol, $K_S = $ [OCH$_3^-$] [H$^+$], is 1.0×10^{-17} M^2. The rate of formation of the α-methoxycarboxylic acid was followed by observing the production of bromide ion. Products of the reaction were isolated and their optical purity was determined. The reactions were studied under pseudo-first-order conditions in buffered solutions, and the following data was obtained:

[H$^+$] (M)	k_ψ (sec^{-1})	Reaction stereochemistry
1.00×10^{-3}	1.01×10^{-8}	100% inversion
1.00×10^{-4}	1.11×10^{-8}	82% net inversion (ie: 91% inversion and 9% retention.)
1.00×10^{-5}	2.1×10^{-8}	4.0% net inversion

1.00×10^{-6}	1.2×10^{-7}	66% net retention
1.00×10^{-7}	1.1×10^{-6}	80% net retention
1.00×10^{-8}	1.0×10^{-5}	82.6% net retention
1.00×10^{-9}	5.5×10^{-5}	82.6% net retention
1.00×10^{-10}	1.0×10^{-4}	80% net retention
1.00×10^{-11}	1.2×10^{-4}	66% net retention
1.00×10^{-12}	2.1×10^{-4}	4.0% net inversion
1.00×10^{-13}	1.1×10^{-3}	82% net inversion

Propose a mechanism for the reaction which is consistent with all of the data, and evaluate as many of the rate constants in the scheme as possible.

HINT: Consider the possibility of a reactive intermediate with the structure:

REFERENCES

1. J. Hine, J. Org. Chem., 31, 1236 (1966). This article contains a discussion of factors which may lead to the formation of the least stable isomers in some reactions.

2. L. P. Hammett, Physical Organic Chemistry, 2nd ed., McGraw-Hill Book Co., Inc., New York, New York, 1970, Chap. 7. The author presents a very thorough discussion of salt effects on rates of reactions and lists primary references.

3. A. J. Parker, Chem. Revs., 69, 1 (1969). This article contains an unusually clear discussion of solvent effects on activity coefficients.

4. J. H. Hildebrand and J. H. Scott, The Solubility of Nonelectrolytes, Dover Publications, Inc., New York, New York, 1964, Chap. 5.

5. E. Grunwald and S. Winstein, J. Amer. Chem. Soc., 70, 846 (1948).

6. E. M. Kosower, J. Amer. Chem. Soc., 80, 3253 (1958).

7. C. G. Swain, R. B. Moseley, and D. E. Bown, J. Amer. Chem. Soc., 77, 3731 (1955).

8. T. W. Bentley, F. L. Schadt, and P. v. R. Schleyer, J. Amer. Chem. Soc., 94, 992 (1972).

9. B. Bensley and G. Kohnstam, J. Chem. Soc., 1955, 3408.

10. J. M. Harris, D. J. Raber, and P. v. R. Schleyer, in Ions and Ion-Pairs in Organic Reactions, Vol. II (M. Szwarc, ed.), Wiley Inter-science, Inc., New York, New York, 1974.

11. S. Winstein, P. E. Klinedienst, and G. C. Robinson, J. Amer. Chem. Soc., 83, 885 (1961).

12. J. Hine, Physical Organic Chemistry, McGraw-Hill Book Co., Inc., New York, New York, 1962, Chap. 6, This chapter contains further examples and primary references.

13. H. R. Snyder and J. H. Brewster, J. Amer. Chem. Soc., 71, 291 (1949); S. Siegel and A. F. Graefe, J. Amer. Chem. Soc., 75, 4521 (1953).

14. C. D. Ritchie and G. A. Chappel, J. Amer. Chem. Soc., 92, 1819 (1969); A. Dedieu and A. Viellard, Reaction Transition States, Gordon and Breach, Inc., London, England, 1972.

Chapter 3

COMBINATIONS OF KINETIC, STEREOCHEMICAL, AND PRODUCT STUDIES

In the first two chapters, we have discussed rather straightforward kinetic techniques for gaining information concerning reaction mechanisms. In the present chapter, we shall discuss how several techniques, involving kinetic, stereochemical, and product studies, can be combined to furnish powerful probes of the details of mechanisms. The utility of these probes is best illustrated by application to specific examples, and we shall continue to develop the S_N1 and S_N2 mechanisms in this context.

The simple mechanisms of the S_N1 and S_N2 reactions which we have written in the previous sections provided a basis for qualitative understanding of these reactions for many years. We have already mentioned, however, certain details of the S_N1 reactions that are difficult to reconcile with the simple two-step mechanism involving a carbonium ion intermediate. In particular, the incomplete racemization observed in these reactions is not expected on the basis of the postulated "free" carbonium ion intermediate, and the unusual salt-dependent mass law effect appears inexplicable in terms of the simple mechanism.

3-1. INTERMEDIATES IN THE S_N1 REACTION

The first clear indication of the origin of the unexpected stereochemistry of the S_N1 reaction came from a study of the solvolysis of 3-methyl-3-chloro-1-butene in acetic acid solution buffered with acetate ion [1]:

Isomeric allylic acetates

The reaction of pure α,α-dimethylallyl chloride produced isomeric γ,γ-dimethylallyl chloride simultaneously with a mixture of the acetate products.

Since the γ,γ-dimethyl isomer reacts much more slowly than the originally added α,α-isomer, the rate of its formation could be determined. If the reaction were a simple S_N1 type, the formation of the γ,γ-isomer could be postulated to arise from trapping of the intermediate carbonium ion with chloride ion. If this were the case, the rate of formation of the isomeric chloride should be increased by addition of chloride ion to the solution. It was observed, however, that added chloride ion had no effect on the rates. It was therefore postulated that the reactions were proceeding through an intermediate in which the carbon-chlorine bond is broken, but which has the chloride ion still in close proximity to the allylic cation moiety. This intermediate, then, is an ion pair, and not a "free" carbonium ion. The collapse of the ion pair to form the γ,γ-dimethylallyl chloride was termed "internal return."

The role of ion pair intermediates in the S_N1 type of reaction was elaborated through studies of salt effects on acetolysis reactions of optically active substrates [2].

The reaction of optically active threo-3-p-anisyl-2-butyl p-bromobenzenesulfonate (III):

III

in acetic acid was followed by observing both the rate of loss of optical activity and the rate of production of p-bromobenzenesulfonic acid. The first-order rate constant for loss of optical activity, k_α, and that for formation of p-bromobenzenesulfonic acid, k_t, were found to vary with changes in concentration of added $LiClO_4$ as shown in Table 3-1. In the absence of $LiClO_4$, we note that k_α/k_t of much greater than 1 means that substrate is being racemized faster than it is going on to product. When lithium brosylate was added to the solution with no lithium perchlorate present, only a normal salt effect on both k_α and k_t was observed. This rules out the possibility that racemization of substrate arises from a "free" carbonium ion intermediate.

The effect of added lithium perchlorate on k_α is that of a normal salt, correlated by the equation

$$k_\alpha = k_\alpha^\circ (1 + 14[LiClO_4]) \tag{3-1}$$

TABLE 3-1

Acetolysis of Threo-3-p-anisyl-2-butyl brosylate [2] at 25°C[a]

[LiClO$_4$] (M)	10^5k_α (sec^{-1})	10^5k_t (sec^{-1})	k_α/kt
0.000	8.06	1.96	4.1
1.0 x 10^{-3}	-	2.68	-
5.0 x 10^{-3}	-	4.69	-
1.0 x 10^{-2}	8.73	5.70	1.54
3.0 x 10^{-2}	11.1	8.40	1.32
6.0 x 10^{-2}	15.6	11.5	1.36
1.0 x 10^{-1}	19.8	16.1	1.23

[a] Brosylate ≡ p-bromobenzenesulfonate.

The effect of lithium perchlorate on k_t, however, is quite different, giving a sharp increase in k_t with low concentrations of salt and then a more normal salt effect at higher concentrations. The ratio k_α/k_t decreases dramatically with small amounts of lithium perchlorate and then remains constant at 1.3 from 3 x 10^{-2} to 10^{-1} M lithium perchlorate.

These effects of lithium perchlorate on k_t can be understood if we postulate that some intermediate of the solvolysis reaction, which can normally return to give racemized substrate, is "trapped" by the lithium perchlorate to give another intermediate which cannot return efficiently to substrate, but goes on to solvolysis product. The leveling off of k_t occurs when the concentration of lithium perchlorate is high enough that all of the intermediate is trapped.

Further evidence for this proposal is furnished by the observation of a salt-induced mass law effect. We have already noted above that the solvolysis reaction in the absence of added lithium perchlorate shows no mass law effect on addition of lithium brosylate. When lithium perchlorate is present, however, a mass law effect is observed [3]. Thus, an intermediate must be formed in the presence of the perchlorate which can react with brosylate ion to reform substrate.

We note further that even at very high concentrations of lithium perchlorate where k_α/k_t is constant, this rate ratio is still considerably greater than 1. Therefore, substrate is still being racemized even when all of intermediate trappable by perchlorate is being diverted to products. Moreover, since k_α/k_t is insensitive to salt effects at this point, the

transition state for the racemization reaction must be showing the same
activity coefficient behavior as does the transition state for the solvolysis
reaction. These observations lead to the postulation of another intermediate, which must precede the one which can be trapped by perchlorate.

The full mechanistic scheme proposed to account for the experimental
observations is the following:

1. $RX \rightleftharpoons (R^+X^-)$ (an "intimate ion pair")

2. $(R^+X^-) \rightleftharpoons (R^+//X^-)$ (a "solvent-separated ion pair")

3. $(R^+//X^-) + Y^- \rightleftharpoons (R^+//Y^-) + X^-$ (anion exchange)

4. $(R^+//X^-) + \text{solvent} \longrightarrow \text{product}$

5. $(R^+//Y^-) + \text{solvent} \longrightarrow \text{product}$

X is p-bromobenzenesulfonate and Y is perchlorate in the specific case
discussed here.

One other piece of evidence for this mechanism was furnished by noting
that if Y^- were an anion which could form a stable molecule, RY, then the
addition of LiY to the solvolysis reaction should give rate effects similar
to those observed with $LiClO_4$, but that a substantial amount of RY should
be formed along with the solvolysis product. The addition of lithium bromide to the solvolysis reaction gave the predicted result [4].

It is rather curious that the intimate ion pair is not trapped by perchlorate ion. In studies of the solvolysis of other substrates, even more effective trapping agents, such as azide ion, appear incapable of intercepting the
intimate ion pairs. (Contrast this with the later discussion of S_N2 reactions.)

A great deal of work has gone into attempts to characterize the reaction
patterns of the ion pairs involved in S_N1 reactions. It was originally hoped
that the relatively simple studies of racemization and solvolysis would
allow one quantitatively to assess the partitioning of the ion pairs between
product formation and return to starting material. In order to check this
possibility, another technique for detecting ion pairs was introduced: the
scrambling of an isotope of oxygen between the carbonyl and ether positions
of specifically labeled esters [5].

In the reaction of optically active p-chlorobenzhydryl p-nitrobenzoate
specifically labeled at the carbonyl oxygen with ^{18}O [6],

three rate constants can be measured during the solvolysis process. The first-order rate constant for loss of optical activity, k_α, and the first-order rate constant for formation of p-nitrobenzoic acid, k_t, can be determined by techniques which we have already discussed. The rate of equilibration of ^{18}O between the ether and carbonyl positions is a kinetic problem similar to that which we discussed for isotopic exchange reactions.

For the process

the reaction will proceed until the ^{18}O is equally distributed between the two positions. We define a reaction variable y

$$y = 1 - \frac{2[E]}{[C]_0} = \frac{[C]_0 - 2[E]}{[C]_0}$$

(3-2)

which has a value of 1 at t = 0, and a value of 0 at t = ∞. The development of the kinetic expression is straightforward:

$$\frac{dy}{dt} = -\frac{2}{[C]_0}\frac{d[E]}{dt} = -\frac{2}{[C]_0} k([C] - [E])$$

$$= -2k\left[\frac{[C]}{[C]_0} - \frac{[E]}{[C]_0}\right] = -2k\left[\frac{[C]_0 - 2[E]}{[C]_0}\right]$$

$$\frac{dy}{dt} = -2ky \qquad \ln y = -2kt \equiv -k_{eq}t$$

(3-3)

The actual measurement of y involves isolation of unreacted ester at various times, chemical degradation, and finally, mass spectrometric analysis of the degradation products.

The results of the study of the solvolysis of p-chlorobenzhydryl p-nitrobenzoate in aqueous acetone at 99.6°C are shown in Table 3-2. Solvolysis in the presence of added ^{14}C-labeled p-nitrobenzoic acid gave no incorporation of radioactivity into the starting ester, ruling out the possibility that racemization or ^{18}O scrambling is due to return from a carbonium ion intermediate.

The process measured by k_α involves both racemization of starting ester and formation of product. The rate constant for racemization of starting ester, k_{rac}, is, therefore

$$k_{rac} = k_\alpha - k_t$$

(3-4)

TABLE 3-2

Solvolysis of Optically Active, Carbonyl ^{18}O-labeled
p-Chlorobenzhydryl p-Nitrobenzoate at 99.6°C [6][a]

	90% acetone	80% acetone
k_α	2.7×10^{-6}	2.0×10^{-5}
k_t	1.4×10^{-6}	1.2×10^{-5}
k_{eq}	3.5×10^{-6}	1.8×10^{-5}

[a] Values for rate constants are given in sec^{-1}.

The data in Table 3-2 show that k_{eq} is larger than k_{rac} in both solvent systems. If we assume that formation of the intimate ion pair results in the
ether and carbonyl oxygen becoming equivalent, we can calculate the
amount of racemization which occurs in the ion pairs, and also obtain the
partitioning of the ion pairs between return and product formation. In 90%
acetone, 71% of the ion pairs return to ester $[k_{eq}/(k_{eq} + k_t)]$, and of those
returning, 19% return with inversion of configuration ($k_{rac}/2k_{eq}$) and 81%
with retention of configuration. Similarly, in 80% acetone, 60% of the ion
pairs return to ester, 22% with inversion of configuration.
      When the above reaction was studied in 80% acetone with 0.141 M NaN_3
present, it was observed that $k_\alpha = k_t$; scrambling of the ^{18}O in the starting
ester was still observed, however. This result indicates that racemization
of starting ester occurs only at the solvent-separated ion pair stage,
whereas the ^{18}O scrambling occurs at the intimate ion pair stage of the
solvolysis reaction.
      Unfortunately, these latter conclusions cannot be generalized. There
are now several cases known for which k_{rac} is greater than k_{eq}, and some
cases where added azide ion does not stop racemization of starting ester.
Some illustrative data are listed:

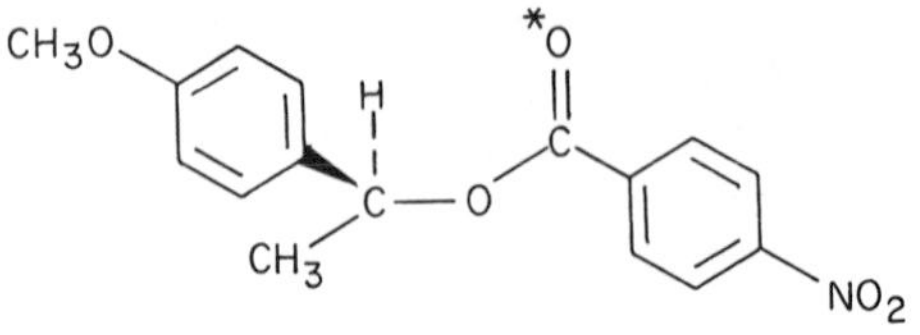

In aqueous acetone with no azide, $k_{eq} = 2k_{rac} = k_t$; with 0.14 M NaN$_3$, $k_{eq} = 2k_{rac} = 0.5k_t$ [7].

In 90% acetone at 48.8°C, $k_t = 1.0 \times 10^{-7}$ sec^{-1}, $k_{eq} = 2.9 \times 10^{-7}$ sec^{-1}, $k_{rac} = 5.0 \times 10^{-8}$ sec^{-1} [8].

In 90% acetone at 48.8°C, $k_t = 3.6 \times 10^{-5}$ sec^{-1}, $k_{eq} = 8.6 \times 10^{-5}$, $k_{rac} = 2.4 \times 10^{-5}$ [8].

In acetic acid, $k_{rac} = 2k_{eq}$ [9].

Studies of the reactions of allylic systems give even more detailed information concerning the nature of the ion pair intermediates in solvolysis reactions. A particularly informative study of the solvolysis of both allylic isomers of trans-α,γ-phenylmethylallyl p-nitrobenzoate has been reported [10]. The reaction scheme considered is as follows:

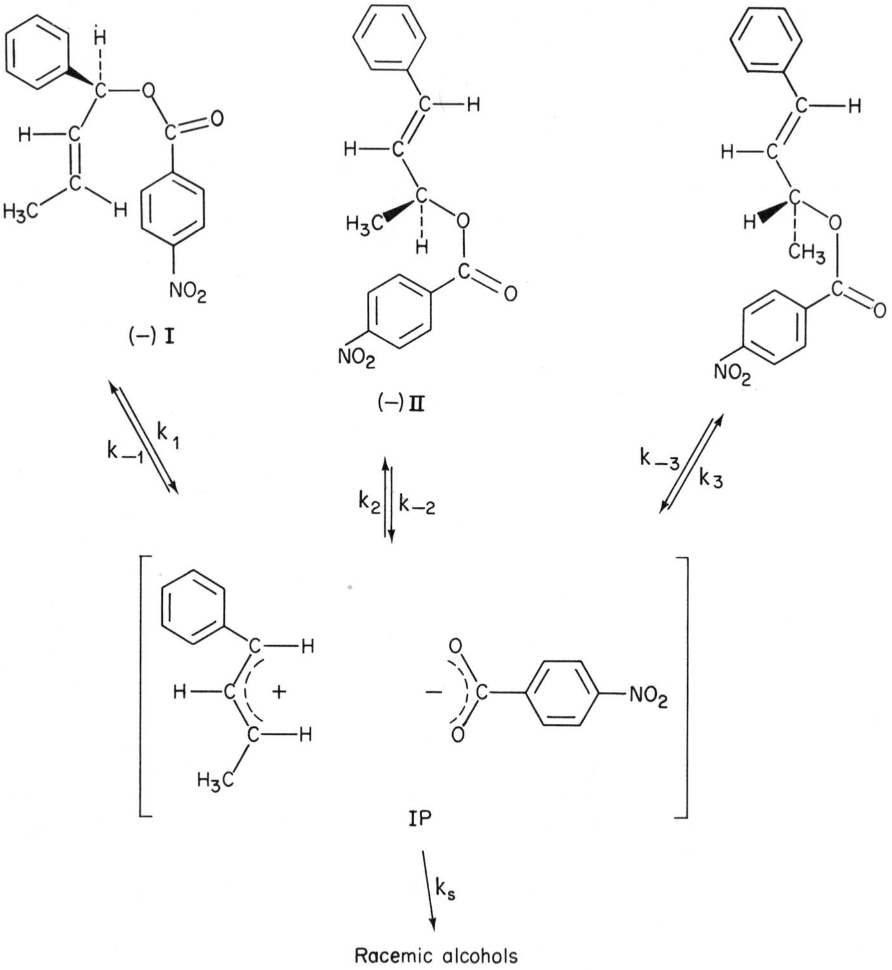

(−) I

(−) II

k_1 / k_{-1}

k_2 / k_{-2}

k_{-3} / k_3

IP

k_s

Racemic alcohols

During the solvolysis reaction, there is no cis-trans isomerization observed. This is consistent with the expectation that the "W" conformation
of the allyl system, as drawn for IP in the reaction scheme, will be
strongly favored by stereoelectronic factors. Examination of models indicates that overlap of the carbon-oxygen bond orbital with the allyl π-
system orbitals can occur with minimum steric crowding in a transition
state which has begun to adopt the "W" conformation:

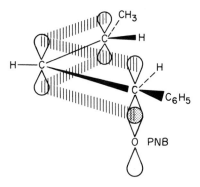

We will, therefore, limit our discussion to intermediates in which this conformation is preserved.

If the p-nitrobenzoate moiety remains on the same side of the allyl cation in the ion pair, allylic rearrangement will interconvert (-)I and (-)II. If the anionic moiety moves from one side to the other of the cation during allylic rearrangement, (-)I and (+)II will be interconverted. Note also that the isomers II, which have the phenyl ring in conjugation with the double bond, will be more stable (i.e., will solvolyze more slowly) than I. Thus, if k_2 or k_3 are comparable in magnitude to k_s, the solvolysis of I will lead to a buildup of II in the solution. In fact, II solvolyzes so slowly compared to I that it may be treated as a stable product during the solvolysis of I (the k_t of I is ca. 300 times the k_t of II.).

The reaction of carbonyl ^{18}O-labeled (-)I produced 71.5% of II and 28.5% of racemic alcohols in aqueous acetone. Optically pure (-)I produced 67.5% (-)II and 32.5% (+)II. In the II produced, 45% of the ^{18}O label was found in the ether position. According to the above reaction scheme, then, the product ratio yields $(k_2 + k_3)/k_s = 2.5$; and the stereochemical result gives $k_2/k_3 = 2.1$. This means that the ion pair collapses 2.5 times faster than it reacts with solvent, and the p-nitrobenzoate moves from one side of the cation to the other only half as fast as it collapses. The ^{18}O results show that the two oxygens become equivalent at a rate only 10% slower than collapse of the ion pair.

In this discussion, we have assumed that I and II have a common ion pair, IP, in their reactions; that is, that the scheme given above is sufficient. We can further check this by solvolyzing optically active (-)II with specific ^{18}O carbonyl label and observing the rate of racemization of starting substrate with rate constant k_{rac}, the rate of oxygen equilibration with rate constant k_{eq}, and the rate of solvolysis with rate constant k_t. From the reaction scheme given, the measured rate constants can be related to rate constants for the various paths:

$$k_{rac} = 2 \; \frac{k_{-2}k_3}{k_3 + k_2 + k_s} \tag{3-5}$$

$$k_{eq} = 0.90 k_{-2} \; \frac{k_2 + k_3}{k_2 + k_3 + k_s} \tag{3-6}$$

$$k_t = \frac{k_{-2}k_s}{k_2 + k_3 + k_s} \tag{3-7}$$

The factor of 0.9 in Eq. (3-6) comes from the observation that 90% oxygen scrambling occurs in the conversion of I to II.

Combining these equations, we can obtain expressions for ratios of the observed rate constants in terms of the rate constant ratios in the reaction scheme which were evaluated above from the experiment starting with (-)I.

$$\frac{k_{rac}}{k_{eq}} = \frac{2.2k_3}{k_3 + k_2} = 0.71 \tag{3-8}$$

$$\frac{k_{eq}}{k_t} = \frac{0.9(k_2 + k_3)}{k_s} = 2.2 \tag{3-9}$$

The actual measurements of k_{rac}, k_{eq}, and k_t, starting with optically pure, specifically labeled (-)II, give $k_{rac}/k_{eq} = 0.65$, and $k_{eq}/k_t = 2.1$. These values are within experimental error of those calculated in Eqs. (3-8) and (3-9), lending strong credibility to the proposed reaction scheme.

A question not answered by the above data is whether the amount of oxygen scrambling in (-)II and (+)II produced from (-)I is the same. We might suspect that as the anion moves from one side to the other of the cation, the oxygen becomes completely scrambled; whereas if the anion stays on the same side of the cation, there might be a preference of the original carbonyl carbon to become the ether carbon in the allylic isomer.

In the solvolysis reaction of cis-5-methyl-2-cyclohexenyl p-nitrobenzo-ate, k_t, k_{rac}, and k_{eq} were measured as in the above case, but it was also possible to isolate the unreacted starting ester, resolve it into enantiomers, and determine the ^{18}O distribution in each enantiomer [11]. The course of the reaction is indicated in the following scheme:

Cis-trans isomerization does not occur in the solvolysis reaction, presumably because of the conformational stability of the cyclic allylic cation. The allylic isomerization, then, leads only to formation of enantiomers of the original ester. If the allylic shift proceeds by a "walk" of the carboxyl group, in which the carbonyl carbon of (+)IIIc bonds to the allylic position, then (-)IIIe will be formed. If the allylic shift occurs by a "slide" of the carboxyl group, where the ether oxygen of (+)IIIc moves to the allylic position, then (-)IIIc will be formed. If the two oxygens become equivalent in the ion pair, then equal amounts of (-)IIIc and (-)IIIe will be formed.

The experimental study gave $k_{rac} = k_{eq}$, and showed that equal amounts of (-)IIIc and (-)IIIe are formed. Thus, the two allylic carbons and the two carboxyl oxygens have each become equivalent in the ion pair intermediate. This result, however, is not general even for cyclic allylic systems. In the solvolysis of trans-5-methyl-2-cyclohexenyl p-nitrobenzoate, k_{eq} is greater than k_{rac} [12].

A very similar study has been carried out on the solvolysis of threo-3-phenyl-2-butyl tosylate in acetic acid [9]. The rate of racemization and the rate of equilibration of ^{18}O between the sulfonyl and ether position were measured, and it was also possible here to resolve the enantiomers of the starting ester and determine ^{18}O distribution in each enantiomer.

As we shall discuss later, there is good evidence for believing that an intermediate with the following structure:

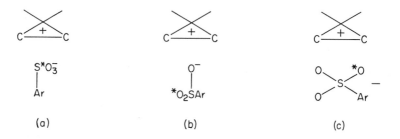

is involved in the reaction of this substrate. Interchange of the positions of the sulfonate and phenyl group through such an intermediate is quite similar to the allylic rearrangements which we have just discussed. The rate study for the reaction showed that k_{rac} is ca. equal to twice k_{eq}, showing that the carbons and oxygens do not become indistinguishable in the ion pair. We may consider three plausible structures for the ion pair intermediate:

(a)   (b)   (c)

Structure (c) would be that involved in a "walk" process, structure (b) that involved in a "slide" process, and structure (a) one with all three oxygens equivalent. None of these three structures alone could give the observed result $k_{eq}/k_{rac} = 0.5$ for reaction of sulfonyl ^{18}O-labeled ester. Structure (a) would give $k_{eq}/k_{rac} = 1$; structure (b) would give $k_{eq}/k_{rac} = 0$; and structure (c) would give $k_{eq}/k_{rac} = 0.75$ (in trace labeling, only one of the sulfonyl oxygens is labeled). You should be able to verify these ratios by kinetic analysis along the lines of Eqs. (3-2) and (3-3). (Hint: set the variable y equal to $1 - 3E/S_0$, where E is the concentration of ether-labeled ester and S_0 is the initial concentration of sulfonyl-labeled ester.)

The observed value of $k_{eq}/k_{rac} = 0.5$ could arise from a 1:1 ratio of structures (a) and (b), or from a 1:2 ratio of structures of (b) and (c). If the latter possibility occurred, the starting enantiomer would show no scrambling of its label, whereas the former possibility would give scrambling of

label in both enantiomers. It is, in fact, possible to calculate the amount
of scrambling in each enantiomer predicted by either of the possibilities.

The measurement of ^{18}O distribution in the ester after partial solvolysis
showed 17% of the label in the ether position of the enantiomer of the starting ester, and appreciable scrambling in the original enantiomer. Thus,
this reaction proceeds through intermediates with structures (a) and (b).

The entire discussion above in terms of the structures of ion pair intermediates could just as properly be given in terms of competing rates of
various processes at the ion pair stage. We may visualize the formation of
an ion pair in which the carbon and oxygen which were bonded in the substrate remain in close contact with each other. Collapse of this species
would simply form original reactant without oxygen scrambling, racemization, or rearrangement. Such a process is termed "invisible return" because none of the techniques which we have discussed can detect it. This
initially formed ion pair can then undergo processes in which the anion moves
relative to the cation, resulting in rearrangements or racemizations, or
the oxygens of the anions may exchange positions with each other, resulting
in ^{18}O scrambling. All of these processes will compete with collapse of
the ion pair. In terms of these processes, the "structures" that we have
discussed are time-average structures over the lifetime of the ion pair.

If one is to detect the process that we have termed "invisible return,"
it is necessary to generate the ion pair intermediate from some reactant
other than the ester whose solvolysis is studied. There have been several
attempts along these lines. One such experiment involved the reaction of
an olefin with p-bromobenzenesulfonic acid, which presumably proceeds
through protonation of the olefin as the initial step [13].

Propene in trifluoroacetic acid solvent forms 2-propyl trifluoroacetate
with a half-life of ca. 300 min under the conditions used in the experiments.
Propene in the presence of 0.10 M p-bromobenzenesulfonic acid in trifluoroacetic acid, under the same conditions, gives only 2-propyl brosylate
if the product is isolated a few minutes after mixing the reagents. The 2-
propyl brosylate solvolyzes slowly ($t_{1/2}$ = 182 min) under these conditions.
In the trifluoroacetic acid solution, p-bromobenzenesulfonic acid is not
dissociated to any appreciable extent. It was argued that protonation of
propene by p-bromobenzenesulfonic acid led to formation of an intimate ion
pair of 2-propyl cation and brosylate anion. The above experiments then
indicate that this ion pair collapses to the 2-propyl brosylate much faster
than it reacts with solvent. If one accepts these arguments, then it follows
that there is extensive "invisible return" in the solvolysis of 2-propyl
brosylate in trifluoroacetic acid solution. The weakness in this argument
is that it is possible that the reaction of propene with p-bromobenzenesulfonic acid follows a concerted mechanism without intermediacy of the intimate ion pair.

Another study involved the formation of a benzhydryl benzoate ion pair
from the reaction of benzoic acid with diphenyldiazomethane in ethanol

solution [14]. Independent experiments indicate the following mechanism for this reaction:

$$(C_6H_5)_2C{=}N_2 + C_6H_5COOH \longrightarrow [(C_6H_5)_2\overset{H}{\underset{|}{C}}{-}N_2^{+\ -}O_2CC_6H_5]$$

$$\xrightarrow{\text{fast}} [(C_6H_5)_2\overset{H}{\underset{|}{C}}{^{+\ -}}O_2CC_6H_5] + N_2$$

$$k_s \diagup \qquad\qquad \diagdown k_r$$

$$(C_6H_5)_2CHOEt \qquad\qquad (C_6H_5)_2CHO_2CC_6H_5$$

The rate of ethanolysis of benzhydryl benzoate under the experimental conditions is much slower than the rate of the diphenyldiazomethane reaction, so that if products are isolated quickly from the latter reaction, the ratio of products gives a direct measure of k_s/k_r. The experiment gave $k_s/k_r =$ 1.7. This is the same value that is found for k_{eq}/k_t in the solvolysis of carbonyl ^{18}O-labeled benzhydryl benzoate under the same conditions. The agreement in values indicates that there is no invisible return in the solvolysis of the labeled ester, and that the same ion pair intermediates are formed in the solvolysis of benzhydryl benzoate and in the reaction of diphenyldiazomethane with benzoic acid. One should, however, have proper respect for coincidences in science. It is possible that different ion pair intermediates are involved in the two reactions, that there is appreciable invisible return in the ethanolysis of benzhydryl benzoate, and that these two factors operate in such a way that the equality of the above rate ratios results.

3-2. THE QUESTION OF INTERMEDIATES IN THE S_N2 REACTION [15]

The reaction of optically active 2-octyl brosylate with azide ion in 75:25 dioxane:water solution produces 2-octyl azide with 100% inversion of configuration. The rate of reaction is first-order with respect to azide ion concentration. The very similar substrate, 2-octyl mesylate (methanesulfonate), in 25:75 dioxane:water solution reacts with azide ion, also producing 2-octyl azide with 100% inversion of configuration, but the rate of reaction is independent of azide ion concentration. In 70:30 dioxane:water the reaction is close to second-order (first-order with respect to azide ion) at low concentrations of azide ion, but at higher concentrations the reaction becomes independent of azide ion concentration.

It was suggested that these results are most concisely understood in terms of a mechanism similar to that of an S_N1 reaction, but with a "tight ion pair" intermediate which can react with azide ion giving inversion of configuration. The changes in kinetic order observed would arise from changes in the rate–determining step. The alternative explanation of the kinetics would be some combination of the classical S_N2 and S_N1 mechanisms operating simultaneously.

These two possibilities, simultaneous S_N1 and S_N2, or single pathway mechanisms, can be distinguished by a study of product distribution and kinetics, which avoids some of the problems of salt effects in the interpretation of kinetic data alone. Consider first the scheme for simultaneous operation of an S_N1 and S_N2 mechanism:

$$S_N1 \begin{cases} R\text{-}X \underset{k_{-1}}{\overset{k_1}{\rightleftharpoons}} R^+ + X^- \\[2mm] R^+ + H_2O \xrightarrow{k_s} ROH + H^+ \\[2mm] R^+ + N_3^- \xrightarrow{k_N} RN_3 \end{cases}$$

$$S_N2 \begin{cases} RX + H_2O \xrightarrow{k_{2s}} ROH + H^+ \\[2mm] RX + N_3^- \xrightarrow{k_{2N}} RN_3 \end{cases}$$

Application of the Bodenstein approximation to R^+ and use of an excess of azide ion to give pseudo–first–order kinetics gives

$$k_\Psi = \frac{k_1[k_N(N_3^-) + k_s]}{k_{-1} + k_N(N_3^-) + k_s} + k_{2N}(N_3^-) + k_{2s} \tag{3-10}$$

$$m \equiv \frac{[RN_3]_\infty}{[ROH]_\infty} \frac{1}{(N_3^-)} = \frac{k_1 k_N + k_{2N}[k_{-1} + k_N(N_3^-) + k_s]}{k_1 k_s + k_{2s}[k_{-1} + k_N(N_3^-) + k_s]} \tag{3-11}$$

where the terms in (N_3^-) are actually activities of the azide ion. If the activity of the azide ion increased only slightly with increasing concentration of sodium azide, the very slight dependence of the pseudo–first–order rate constant on azide ion concentration observed could be explained. If this were the cause of the kinetic behavior, however, one would expect the product ratio to show a corresponding behavior. The experimental data presented in Table 3-3 show that m, defined in Eq. (3-11), in fact remains constant indicating that the azide ion activity is closely following concentration of sodium azide.

The possibility that the dependence of k_Ψ on sodium azide concentration arises from a "normal" salt effect which decreases k as salt concentration increases, coupled with the kinetic dependence predicted by Eq. (3-10), was considered unlikely [17] since studies of the effect of added $LiClO_4$, $NaNO_3$, and $NaBr$ on the rate of the solvolysis reaction showed increasing

TABLE 3-3

Reaction of 2-Octyl Mesylate in 30:70 Dioxane:Water [16]

[NaN3] (M)	Percent RN3	m	k_ψ (sec^{-1})	k_{corr}^a (sec^{-1})	k_{calc}^b (sec^{-1})
0	0	–	1.74 x 10^{-4}	1.74 x 10^{-4}	1.74 x 10^{-4}
0.054	38.4	11.5	2.43 x 10^{-4}	2.30 x 10^{-4}	2.28 x 10^{-4}
0.057	39.3	11.4	2.26 x 10^{-4}	2.14 x 10^{-4}	2.30 x 10^{-4}
0.098	52.5	11.3	2.67 x 10^{-4}	2.42 x 10^{-4}	2.66 x 10^{-4}
0.152	54.5	7.9	3.65 x 10^{-4}	3.16 x 10^{-4}	3.00 x 10^{-4}
0.199	64.0	8.9	3.73 x 10^{-4}	3.10 x 10^{-4}	3.26 x 10^{-4}
0.258	69.3	8.8	4.71 x 10^{-4}	3.73 x 10^{-4}	3.53 x 10^{-4}
0.311	74.6	9.5	4.91 x 10^{-4}	3.71 x 10^{-4}	3.62 x 10^{-4}

a Eq. (3-12).
b Eq. (3-18).

rate with increasing salt concentration. The effect of these salts was correlated by the Eq. (3-12),

$$k_0 = k_0^0 [1 + b(salt)] \qquad\qquad (3-12)$$

with b = 1.04 for LiClO$_4$, 0.73 for NaNO$_3$, and 0.73 for NaBr. The k_{corr} values shown in Table 3-3 are calculated from Eq. (3-12) using a value of b = 1.04, and are the experimentally observed pseudo-first-order rate constants corrected to zero salt concentration.

Consider now the scheme for a single pathway mechanism:

$$RX \underset{k_{-1}}{\overset{k_1}{\rightleftharpoons}} R^+X^-$$

$$R^+X^- + H_2O \xrightarrow{k_S} ROH + H^+$$

$$R^+X^- + N_3^- \xrightarrow{k_N} RN_3$$

Again using Bodenstein approximation and pseudo-first-order conditions, we can derive the following equations:

$$k_\Psi = \frac{k_1[k_N(N_3^-) + k_s]}{k_{-1} + k_s + k_N(N_3^-)} \qquad (3\text{-}13)$$

$$m \equiv \frac{[RN_3]_\infty}{[ROH]_\infty} \frac{1}{(N_3^-)} = k_N/k_s \qquad (3\text{-}14)$$

Combination of these last two equations allow us to write a direct relationship between product ratio and rate constant.

Defining k_0 as the pseudo-first-order rate constant for the reaction in the absence of azide ion gives

$$k_0 = \frac{k_1 k_s}{k_{-1} + k_s} \qquad (3\text{-}15)$$

Note that we can take account of "normal" salt effects on k_0 by the use of Eq. (3-12) with the value of k_0° obtained in the absence of all salts.

Further defining

$$x = \frac{k_{-1}}{k_s} \qquad (3\text{-}16)$$

then

$$k_0 = \frac{k_1}{(x + 1)} \qquad (3\text{-}17)$$

Combination with Eqs. (3-13) and (3-14) gives

$$\frac{k_\Psi}{k_0} = \frac{(x + 1)\,[1 + m(N_3^-)]}{x + 1 + m(N_3^-)} \qquad (3\text{-}18)$$

Note particularly that x is independent of both the identity and the concentration of the trapping reagent. Since m is found experimentally to be independent of salt concentration, we may reasonably assume that x is also independent of salt concentration. With this assumption, we can test whether Eq. (3-18) is capable of accomodating the data shown in Table 3-3. The data in the last column of the table, k_{calc}, are calculated from Eq. (3-18) with $x = 2.59$ and $m = 9.04$, the average of the experimental values. In all cases, the calculated values are within experimental error of the salt-effect-corrected observed values k_{corr} lending strong support to the postulated single pathway mechanism. The value of x in these experiments is treated as an adjustable parameter.

An even more convincing demonstration of the applicability of the single pathway mechanism was provided by a study of the reactions of α-phenylethyl bromide and of α-p-tolylethyl chloride with both azide and thiocyanate ions in ethanol solution [18]. The use of two trapping reagents removes the adjustable parameter x from Eq. (3-18). A study analogous to that discussed above using azide ion as the trapping agent allows evaluation of x to fit the kinetic data best. Since x is independent of the identity of the trapping agent, however, the kinetics of the reaction in the presence of thiocyanate must be accommodated by the same value of x. Since the value of m for thiocyanate is determined independently of the kinetics, Eq. (3-18), with no adjustable parameters, must fit the kinetic data for the thiocyanate reaction. For both α-phenylethyl bromide and α-p-tolylethyl chloride, the agreement between calculated and experimental rate constants is excellent.

At the present time the arguments in favor of a tight ion pair intermediate in the S_N2 reaction of secondary alkyl halides, tosylates, etc., appear quite strong. There is considerable controversy, however, as to whether such an intermediate is involved in the reactions of primary systems, or even for all secondary systems [17]. Considerations of the stabilities of primary carbonium ions indicate that the energy of an ion pair intermediate in the solvolyses of primary systems would be prohibitively high [19].

3-3. NEIGHBORING GROUP PARTICIPATION

Another type of intermediate is encountered in solvolysis reactions in cases where the solvolyzing molecule contains, in addition to the leaving group, another group which can act as an intramolecular nucleophile:

The intermediate can then undergo further reaction with solvent to produce the solvolysis product:

Such intermediates result from an intramolecular S_N2 attack, and are frequently followed by an S_N2 displacement by solvent, resulting in a double inversion of configuration, hence retention, at the reacting carbon. In many cases such intermediates can also lead to rearrangements:

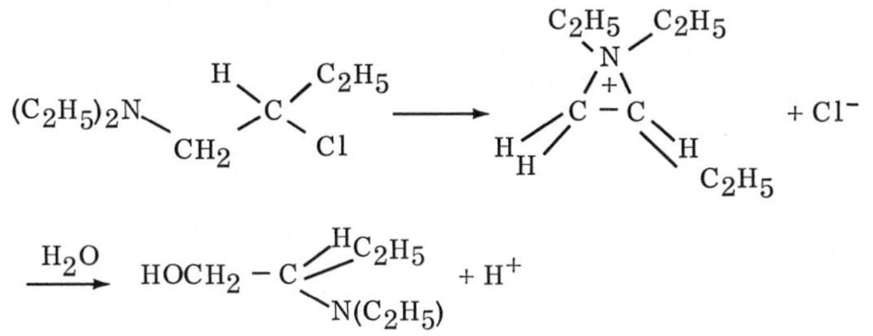

When two asymmetric centers are involved in such a rearrangement, inversion at both centers frequently occurs. A particularly illustrative example is furnished by reaction of 3-hydroxy-2-bromobutane with HBr [20]:

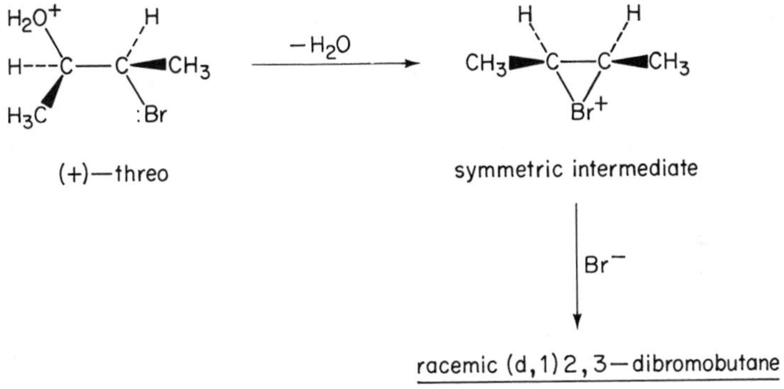

(+)—threo

symmetric intermediate

Br^-

racemic (d,1) 2,3—dibromobutane

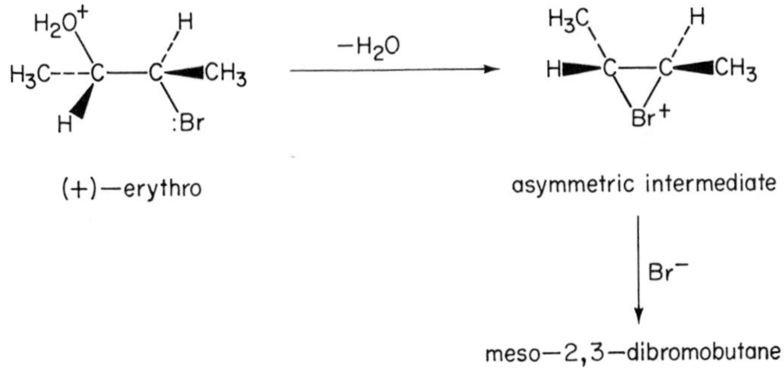

(+)—erythro

asymmetric intermediate

Br^-

meso—2,3—dibromobutane

Problem 10 at the end of Chapter 2 presented data for another case of neighboring group participation; that of carboxylate group participation to form an α-lactone intermediate in the reaction of α-bromopropionic acid. In this case, the data shows clearly that the conjugate base exhibits neighboring group participation while the conjugate acid does not.

It should be obvious that neighboring group participation will occur only if the neighboring group can compete successfully with external nucleophile (usually solvent) in the initial displacement, and that if the intermediate is not detected in the reaction, it must react with external nucleophile faster than does the original reactant. These conditions require that reactions proceeding completely by way of neighboring group participation must have rates greater than that of the normal S_N2 reaction. One of the diagnostics for neighboring group participation, then, is an abnormally fast rate of reaction. The problem here lies in the question of what is to be considered abnormally fast. What we mean, of course, is fast relative to the same molecule without neighboring group participation, but it is not possible to observe the latter process. In general, it is known that electron-withdrawing groups decrease the rates of both S_N1 and S_N2 reactions. These same types of groups are often involved in neighboring group participation. In such cases, then, neighboring group participation is indicated if the group does not decrease the rate of reaction of a parent compound as much as normal. In the next Chapter, we shall treat this question in a quantitative manner. For the present, we can consider the data in Table 3-4 in a qualitative manner.

TABLE 3-4

Relative Rate Constants for Solvolyses
of G-CH$_2$CH$_2$-OTs [21]

G	k_G (rel)
HOCH$_2$CH$_2\overset{..}{S}$	10^7
$\overset{..}{O}{}^-$	10^{10}
$\overset{..}{N}H_2$	10^4
$\overset{..}{I}$	1.6×10^3
H	(1.00)
$\overset{..}{B}r$	0.4
$H\overset{..}{O}$ or $CH_3\overset{..}{O}$	0.1

The large rate enhancements observed on substitution of the β-hydrogen of ethyl tosylate with the mercaptyl, oxy anion, amino, or iodo groups are certainly strong indications of neighboring group participation. We see, however, that substitution of the hydrogen by Br or by OCH_3 actually causes a decrease in the rate of solvolysis. The main argument in favor of neighboring group participation here is that such participation is observed in other compounds where stereochemical results or studies of ^{14}C specifically labeled compounds can serve as probes. It can also be argued that such strongly electron-withdrawing groups would cause much larger decreases in rate if neighboring group participation were not involved.

The ability of a group to show neighboring group participation will depend on a number of factors other than the identity of the group. For example, the group must be able to attack the reacting carbon from the back side since an intramolecular S_N2 process is involved. The clearest demonstrations of the importance of this conformational factor are found in reactions of cis- and trans-1,2-disubstituted cyclohexane derivatives:

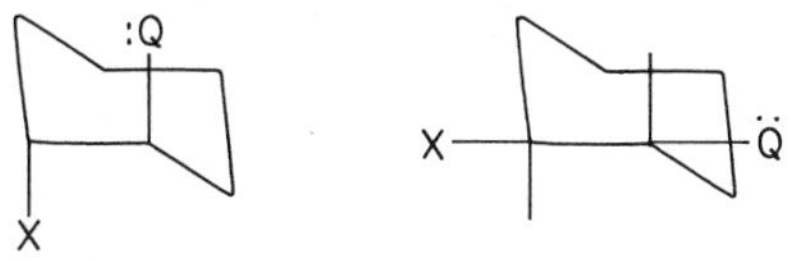

Here, neighboring group participation can occur from the generally favored chair conformation only for the trans-substituted derivatives. Trans-2-acetoxycyclohexyl tosylate undergoes acetolysis at a rate 10^4 greater than that of the cis-isomer [21]. In ethanol, trans-2-acetoxycyclohexyl tosylate forms the orthoester as follows:

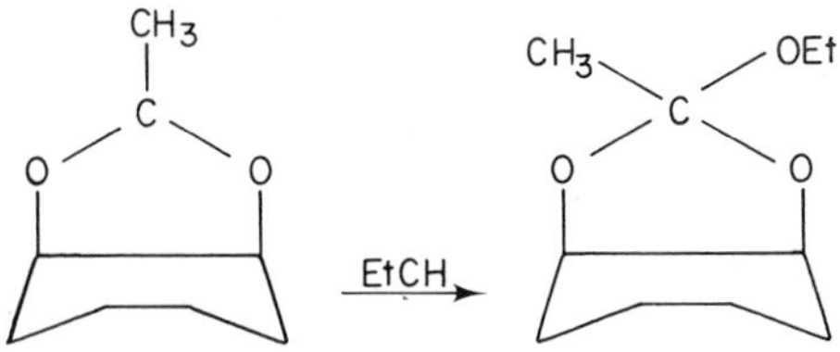

This results from trapping of the bridged ion intermediate [22]. Similarly, trans-2-iodocyclohexyl brosylate undergoes acetolysis to produce trans-2-iodocyclohexyl acetate, and the acetolysis occurs at a rate more than 10^6 greater than that of the cis-isomer [21].

The trans-geometry is, however, not sufficient to allow neighboring group participation unless the ring system involved can allow a trans-coplanar arrangement of the neighboring group and the leaving group. For example, the trans-2-chloro-3-thiophenyl norbornane

shows no evidence of neighboring group participation, presumably because
the boat conformation of the six-membered ring does not allow the necessary trans-coplanar configuration of chlorine and sulfur [23].

The cases of neighboring group participation which we have considered
so far involve formation of either three- or five-membered ring intermediates. Other ring sizes are possible, but four-membered rings or larger
than six-membered rings are not very favorable. In open chain compounds,
the probability of the reactant having a conformation allowing the trans-
coplanar geometry of neighboring group and leaving group decreases as the
chain length increases. This entropy factor, then, decreases the rate constant for the intramolecular process. Ring strain effects operate in the
opposite direction, disfavoring formation of the smaller three- and four-
membered rings. Table 3-5 contains some data illustrating the operation
of these conflicting factors.

Some of the most interesting cases of neighboring group participation
involve carbon as the intramolecular nucleophilic atom. One of the earliest
examples of carbon participation was suggested to account for the stereochemistry of reactions of 3-phenyl-2-butyl derivatives.

TABLE 3-5

Ring Size Effects in Neighboring Group Participation [24]

$$H_2N(CH_2)_nBr \xrightarrow[25\,°C]{H_2O} H_2N(CH_2)_nOH$$

n	k (sec^{-1})
2	6.0×10^{-4}
3	8.3×10^{-6}
4	5.0×10^{-1}
5	8.3×10^{-3}
6	1.7×10^{-5}

The acetolysis of threo-3-phenyl-2-butyl brosylate (optically active) gives 95% racemic threo-3-phenyl-2-butyl acetate and 5% of inverted erythro isomer as the acetate products, along with considerable amounts of elimination product [25]. The actual product distribution is shown in the following scheme:

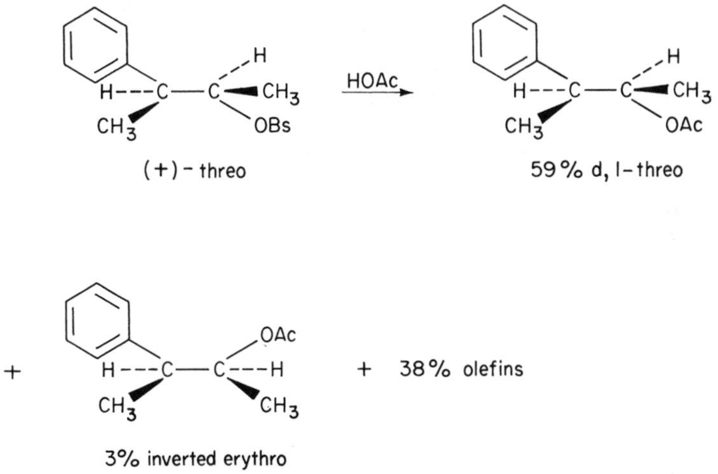

(+)-threo   HOAc →   59% d, l-threo

+   3% inverted erythro   +   38% olefins

We have already discussed the fact that racemization and ^{18}O scrambling occurs in the reactant during acetolysis of the corresponding tosylate. The intermediacy of a symmetrical intermediate in these reactions is particularly suggested by the formation of the d,1-threo acetate product. The formation of erythro acetate, however, suggests that all of the reaction might not proceed through such a symmetrical intermediate. We shall discuss this point later. The acetolysis of threo-3-p-anisyl-2-butyl brosylate gives only the threo-acetate; and in this case, it appears that all of the reaction proceeds through a bridged symmetrical intermediate:

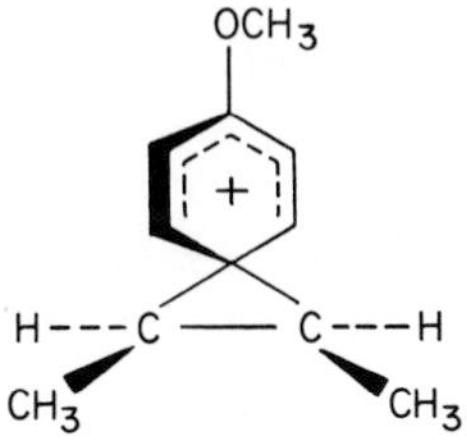

The products obtained from the acetolysis of various threo-3-aryl-2-butyl brosylates are shown in Table 3-6.

If we use the percent of threo-acetate product as a measure of the extent of neighboring group participation by the aryl group in these reactions, we conclude that the p-OCH$_3$ derivative solvolyzes exclusively through the bridged species, and that the p-nitro derivative solvolyzes without aryl participation. We shall see later that this estimate is consistent with other data. The conclusion is that aryl group participation can occur, but is sensitive to substituents on the benzene ring.

Participation of simple double bonds is indicated by rate and product studies of a large number of compounds undergoing solvolyses. Some examples where participation has been suggested are the following:

(NsO = p-nitrobenzenesulfonate)

TABLE 3-6

Products of Acetolysis at 75 °C [25]

X	10^5k_t (sec^{-1})	Percent olefins	Percent erythro-acetate	Percent threo-acetate
p-OCH$_3$	1060	0.3	0	99.7
p-CH$_3$	81.4	12	0	88
m-CH$_3$	28.2	31	1	68
H	18.0	38	3	59
p-Cl	4.53	53	6	39
m-Cl	2.05	76	11	12
m-CF$_3$	1.26	75	14	11
p-NO$_2$	0.495	68	12	1

The species drawn in the brackets above have been considered by many as intermediates in the reactions. There is considerable controversy, however, as to whether these drawings with dashed lines represent resonance hybrids or whether they represent a time averages of rapidly interconverting, but distinct, classical carbonium ions [26]. We shall discuss this distinction in connection with a later discussion of rates of carbonium ion rearrangements, but can note that according to either view, the structures drawn can be used to visualise the reaction courses.

Controversy also exists as to whether saturated carbon is involved in bridged intermediates in some solvolysis reactions [26]. Again, whether or not the bridged species represent stable intermediates, they are useful constructs in visualizing the course of many reactions, such as

We will continue the discussion of the structures of these bridged species after we have developed some concepts of structural effects on reactivity in more common types of reactions.

PROBLEMS

1. On the basis of some structure-reactivity data, it was suggested that 1-p-tolyl-2-propyl tosylate undergoes solvolysis in 80% ethanol by two pathways simultaneously; one path involves aryl participation and a bridged intermediate; the other involves direct nucleophilic attack of solvent on the reactant. From these data, it was estimated that $k_\Delta/k_{2s} \cong 2$, where k_Δ is the rate constant for the aryl participation pathway and k_{2s} is the rate constant for the S_N2 process.
    The solvolysis of the 1-p-tolyl-2-propyl tosylate was then carried out in the presence of various concentrations of sodium azide. The following data on rates and product distributions were obtained under pseudo-first-order conditions:

[NaN$_3$] (M)	k_ψ (sec^{-1})	Percent RN$_3$	Percent ROH	Percent ROEt
0.00	2.40 x 10^{-4}	0	54	46
0.02	3.18 x 10^{-4}	27	39	35
0.04	4.28 x 10^{-4}	46	27	27
0.06	5.19 x 10^{-4}	55	21	24

The solvolysis of the isomeric 2-p-tolyl-1-propyl tosylate was carried out under the same conditions. The products consist predominantly of 1-p-tolyl-2-propyl derivatives, which are believed to arise exclusively from the bridged intermediate - the same one which would be involved for the isomeric tosylate. The following product distributions were observed:

1-Aryl-2-propyl Products from
Solvolysis of 2-Aryl-1-propyl
Tosylate

(NaN$_3$), M	Percent RN$_3$	Percent ROH	Percent ROEt
0.00	0	49	49
0.02	5	43	47
0.04	10	39	41
0.06	12	36	39

Are these data consistent with the proposed reaction scheme?

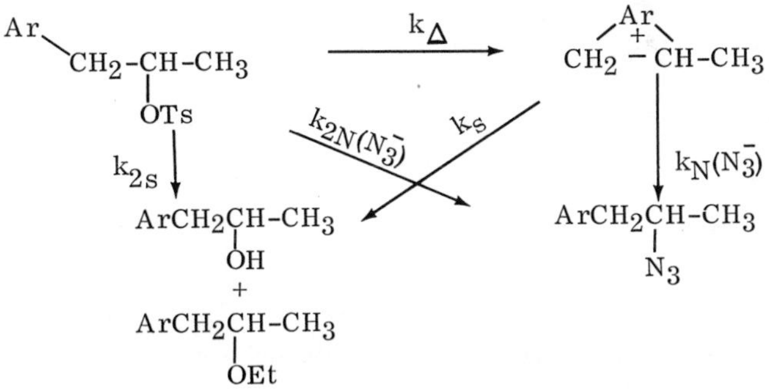

Evaluate as many rate constants as possible from the data. (See Ref. 27.)

2. The solvolysis of either of the two steroidal molecules

or

in methanol solution gives the same products, consisting of two methyl ethers. What are the structures of the products and what intermediate(s) might be involved in the reactions?

3. What product would you expect to result from the acetolysis of each of the following:

a.

b.

c.

4. The reaction of methallyl chloride with hypochlorous acid produces 2,3-dichloro-2-methylpropanol:

$$CH_2 = C\begin{smallmatrix}CH_3\\CH_2Cl\end{smallmatrix} \xrightarrow{HOCl} HOCH_2-C\begin{smallmatrix}CH_3\\Cl\\CH_2Cl\end{smallmatrix}$$

The reaction is believed to occur by initial attack of Cl^+, followed by attack of OH^-. When the reaction was carried out with ^{36}Cl-labeled methallyl chloride, a substantial portion of the label in the product was found in the 2-position. Propose an explanation for this observation.

5. Nitrous acid deamination of alkyl amines is believed to produce carbonium ion intermediates. The product distribution from the deamination reactions, however, is frequently different from that obtained in
solvolyses of corresponding alkyl tosylates. Propose an explanation for
this.

6. The methanolysis of p,p'-dimethoxybenzhydryl mesitoate at 25°C has a
rate constant of 6.35×10^{-4} sec^{-1}, and the same rate constant is observed in the presence of 5×10^{-3} M NaN$_3$.

 A sample of p,p'-dimethoxybenzhydryl mesitoate with tritium substitution in the anisyl rings was prepared

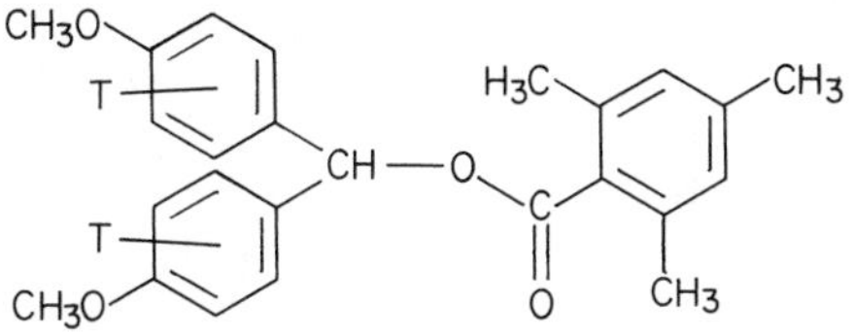

and found to give 8.52×10^6 counts per minute (cpm) per mg of sample
by scintillation counting of β-radioactivity (cpm/mg of sample is proportional to the number of tritium atoms per mg of sample).

 The tritiated ester was then used to study product yields of the methanolysis in the presence of various concentrations of sodium azide,
using the following procedure:

 To 25.0 ml of a methanol solution containing a known concentration
of NaN$_3$ and a buffer to maintain constant pH, 1.15 mg of the tritiated
ester was added. The solution was thermostatted at 25°C and allowed
to stand for 4 hours.

 The solution was then divided into two 10.0 ml portions. In one of
the portions, 50.0 mg of pure (not tritiated) p,p'-dimethoxybenzhydryl
methyl ether was dissolved. In the other portion, 50.0 mg of pure
p,p'-dimethoxybenzhydryl azide was dissolved. The dissolved compounds were then reisolated and each was recrystallized until scintillation counting showed a constant tritium content.

 The following results were obtained in a series of experiments:

NaN$_3$ conc of reaction soln	cpm/mg of reisolated RN$_3$	cpm/mg of reisolated ROCH$_3$
4.0×10^{-4}	2.13×10^4	4.70×10^4
1.0×10^{-4}	7.37×10^3	6.02×10^4
4.0×10^{-5}	3.74×10^3	6.55×10^4
1.0×10^{-5}	1.45×10^3	6.70×10^4

Are all products of the reaction accounted for by the RN_3 and $ROCH_3$? Calculate the product ratios for each of the above runs. What conclusions can be drawn from these results? (See Ref. 28.)

REFERENCES

1. W. G. Young, S. Winstein, and H. L. Goering, J. Amer. Chem. Soc., 73, 1958 (1951).

2. S. Winstein and G. C. Robinson, J. Amer. Chem. Soc., 80, 169 (1958). See Ref. 10 of Chapter 2 for a thorough discussion.

3. S. Winstein, P. E. Klinedienst, and G. C. Robinson, J. Amer. Chem. Soc., 83, 885 (1961).

4. S. Winstein, P. E. Klinedienst, and E. Clippinger, J. Amer. Chem. Soc., 83, 4986 (1961).

5. H. L. Goering, Record Chem. Progr., 21, 109 (1960).

6. H. L. Goering and J. L. Levy, J. Amer. Chem. Soc., 86, 120 (1964).

7. H. L. Goering, R. G. Briody, and G. Sandrock, J. Amer. Chem. Soc., 92, 7401 (1970).

8. H. L. Goering and H. Hopf, J. Amer. Chem. Soc., 93, 1224 (1971).

9. H. L. Goering and R. W. Thies, J. Amer. Chem. Soc., 90, 2968 (1968).

10. H. L. Goering, G. S. Koermer, and E. C. Linsay, J. Amer. Chem. Soc., 93, 1230 (1971).

11. H. L. Goering, J. T. Doi, and K. D. McMichael, J. Amer. Chem. Soc., 86, 1951 (1964).

12. H. L. Goering and J. T. Doi, J. Amer. Chem. Soc., 82, 5850 (1960).

13. V. J. Shiner and P. Dowd, J. Amer. Chem. Soc., 91, 6528 (1969).

14. A. F. Diaz and S. Winstein, J. Amer. Chem. Soc., 88, 1318 (1966).

15. R. A. Sneen, Accts. Chem. Res. 6, 46 (1973). This article contains a review and further references.

16. R. A. Sneen and J. W. Larsen, J. Amer. Chem. Soc., 91, 362 (1969).

17. There is considerable controversy on this point. See D. J. Raber, J. M. Harris, R. E. Hall, and P. v. R. Schleyer, J. Amer. Chem. Soc., 93, 4821 (1971); and for reply, R. A. Sneen and H. M. Robbins, J. Amer. Chem. Soc., 94, 7868 (1972).

18. R. A. Sneen and H. M. Robbins, J. Amer. Chem. Soc., 94, 7868 (1972).

19. M. H. Abraham, J. Chem. Soc., Perkin Trans. II, 1973, 1893.

20. S. Winstein and H. J. Lucas, J. Amer. Chem. Soc., 61, 2845 (1939).

21. E. M. Kosower, Physical Organic Chemistry, John Wiley and Sons, Inc., New York, New York, 1968, p. 105. Original literature references are given.

22. S. Winstein and R. E. Buckles, J. Amer. Chem. Soc., 65, 613 (1943).

23. S. Cristol and R. Arganbright, J. Amer. Chem. Soc., 79, 3441 (1957).

24. For a more detailed discussion and references to the original literature, see Ref. 21 and E. S. Gould, Mechanism and Structure in Organic Chemistry, Holt, Rinehart and Winston, New York, New York, 1959, Chap. 14.

25. H. C. Brown, C. J. Kim, C. J. Lancelot, and P. v. R. Schleyer, J. Amer. Chem. Soc., 92, 5244 (1970).

26. P. D. Bartlett, Non-classical Ions, W. J. Benjamin and Co., New York, New York, 1965.

27. D. J. Raber, J. M. Harris, and P. v. R. Schleyer, J. Amer. Chem. Soc., 93, 4829 (1971).

28. C. D. Ritchie, J. Amer. Chem. Soc., 93, 7324 (1971).

Supplementary Reading

R. W. Alder, R. Baker, and J. M. Brown, Mechanism in Organic Chemistry, John Wiley and Sons, Inc., New York, New York, 1971, pp. 85-111. This book contains a good general discussion of the mechanisms and intermediates involved in the S_N1 reaction.

Chapter 4

STRUCTURE AND REACTIVITY: EMPIRICAL RELATIONSHIPS

4-1. INTRODUCTION

The ultimate goal of the physical organic chemist is to be able to predict
the rate and equilibrium constant for any reaction under any given set of
conditions. In our discussion of salt and solvent effects in Chapter 2, we
saw that one presently settles for the less ambitious goal of being able to
predict how changes in ionic strength or solvent will change rate or equilibrium constants for a given reaction. Similarly, in discussions of structural effects on reactivity, we must presently settle for a goal of being able
to predict the changes in rate or equilibria of reactions caused by small
changes in the structures of reactants.

Since the equilibrium constant for a reaction is directly related to the
standard free energy differences of products and reactants, and since
transition state theory similarly relates rate constants to standard free
energy differences of transition states and reactants, we are interested in
how changes in structure affect the standard free energies of molecules.
At the fundamental level, the problem is one of quantum mechanics and
statistical thermodynamics. In principle, we may solve the Schrödinger
equation to obtain the energy levels of a system, and by application of statistical thermodynamics properly average these energies to obtain the
standard free energy of a system of molecules.

In later chapters, we shall examine these theoretical methods and will
find that in certain cases valuable information can be gained by just such a
procedure as that outlined above. In general, however, the quantitative
results of these methods applied to molecules of any complexity are not
sufficiently accurate to allow useful calculations of either rate or equilibrium constants.

The primary difficulty encountered in applications of the theoretical
calculations is the extreme accuracy required to produce chemically useful
information. An equilibrium constant is changed by a factor of 100 by a
change in ΔG° of only ca. 3 kcal/mole at near room temperature. This
energy is extremely small in comparison with the total energy of a molecule as calculated by the Schrödinger equation. For example, the calculated

total energy of the methane molecule (relative to separated electrons and nuclei) is on the order of 20,000 kcal/mole. Thus, theoretical calculations would have to be accurate to ca. ± 0.01% in order to be useful for calculations of rate or equilibrium constants. At the present time, there does not appear to be any fully justified method to allow the theory to recognize the fact that small changes in structure cause small changes in energy.

Fortunately, in this case, experimental facts reveal a regularity of behavior not anticipated by the basic theories. This regularity of behavior has led to the formulation of empirical rules of surprising generality and accuracy for the correlation and prediction of structural effects on reactivity. At the simplest level, these empirical rules consider the properties of a molecule to be the sums of the properties of the component parts. These rules lead to various additivity schemes for calculation of molecular properties. Further developments then consider deviations from additivity in terms of specific types of interactions between the component parts of a molecule.

4-2. ADDITIVITY SCHEMES [1,2]

The simplest additivity scheme is that of additivity of atomic properties. In this scheme we estimate molecular properties as the sum of the properties of the atoms in the molecule, taking account of the linking of parts only by consideration of the number of atoms bonded to each atom (i.e., the ligancy).

Molecular dimensions, particularly bond lengths, are reasonably well handled in most instances by atomic additivity, in which we assign a covalent radius to each type of atom. In virtually all saturated hydrocarbons, the C-C bond length is very close to 1.54 Å; thus, the covalent bond radius of carbon is 0.77 Å. The halogen-halogen bond length in Cl_2 is 1.98 Å and in I_2 is 2.66 Å. We estimate, then, that the C-Cl bond length of methyl chloride is 1.76 Å, in perfect agreement with the experimental value. Similarly, we estimate the C-I bond length of methyl iodide to be 2.10 Å, again in perfect agreement with experiment.

Inclusion of the ligancy number of the atoms allows for the fact that multiple bonds are generally shorter than single bonds. The C=C bond length is 1.33 Å, and the C≡C bond length is 1.19 Å, for example. Covalent radii for some common atoms are listed in Table 4-1.

This simple additivity scheme works quite well for bond lengths until we begin to consider molecules such as butadiene, in which the central C-C bond is shorter than 1.54 Å, or other molecules where other evidence indicates resonance effects. Even many of these deviations from additivity can be handled if we assign different atomic contributions to different hybridizations of the atoms. For example, if we assign atomic contributions of

TABLE 4-1
Covalent Radii[a] Å [1,2]

Atom	Single bond	Double bond	Triple bond
C	0.77	0.67	0.60
N	0.74	0.62	0.55
O	0.74	0.62	
H	0.37		
F	0.72		
Cl	0.99		
Br	1.14		
I	1.33		

[a] Distances are given in Å.

0.743 Å to the single bond length of $=C-$, and 0.691 Å to the single bond length of $\equiv C-$, we can accurately estimate the single bond lengths even in conjugated molecules.

The additivity scheme for bond lengths, of course, is well known to every student of elementary organic chemistry. What is not quite so well known is that the same simple additivity scheme can be used to estimate thermodynamic properties of molecules. Table 4-2 contains a list of atomic contributions to entropy and heat capacities of molecules. As an example of the use of Table 4-2, consider the molecule $CH_3CH=CHCH_3$. By atomic additivity, the properties of the molecule are those of eight hydrogen atoms, two carbons with ligancy of four, and two carbon atoms with ligancy of three. Therefore, $C_p^\circ = 21.8$ e.u. (i.e., 4 x 3.75 + 8 x 0.85) and $S^\circ = 75.8$ e.u. (i.e., 8 x 21.0 - 2 x 13.5 - 2 x 32.6), which may be compared to the experimental values, $C_p^\circ = 19.0$ e.u. and $S^\circ = 72.1$ e.u. for cis-2-butene, and $C_p^\circ = 21$ e.u., $S^\circ = 70.9$ e.u. for trans-2-butene. Similarly, we calculate for CH_3OH, $C_p^\circ = 10.5$ e.u., $S^\circ = 60.2$ e.u., compared to experimental values $C_p^\circ = 10.5$ e.u. and $S^\circ = 57.3$ e.u.

If one attempts to use the simple atomic additivity scheme for the calculation of ΔH° or ΔG°, unacceptably large deviations are found even with the best parameterization. In order to have a useful additivity scheme for these thermodynamic quantities, we must go to the next level of additivity, that of the additivity of bond properties. Some useful bond properties are shown in Table 4-3.

TABLE 4-2

Atomic Contributions to Heat Capacity and Entropy[a]

Atom	C_p°	S°			
		Ligancy			
		1	2	3	4
H	0.85	21.0			
C	3.75		5.3	-13.5	-32.6
N	3.40	22.9	5.2	-12.1	
O	3.40	25.5	8.8		
F	2.40	25.5			
Cl	3.70	28.4			
Br	4.20	31.3			

[a] Measurements are given in cal/deg-mole at 25°C and 1 atm [1,2].

TABLE 4-3

Bond Additivity Parameters [3]

Bond	Length (Å)	Heat of dissociation (kcal)	Dipole moment (debye)
C–C	1.53	83.1	0
C=C	1.335	145.0	0
C≡C	1.206	198.0	0
C–H	1.12	98.8	0.30
C=C (aromatic)	1.39	–	0
C–O	1.43	84.0	0.86
C=O	1.20	179.0	2.40
N–H	1.03	93.4	1.31
O–H	0.97	117.5	1.53
C–N	1.51	62.0	0.40
C=N	1.29	121.0	0.90
C≡N	1.15	191.0	3.60

The principle of bond additivity, in fact, served for many years as the basis for evaluation of resonance effects. Basically, it was assumed that bond additivity would be exact unless specific resonance or steric effects operated. For example, the fact that heats of hydrogenation of the compounds shown in Table 4-4 varied was taken as evidence of resonance interactions, such as hyperconjugation of the methyl groups. The resonance energies of some compounds calculated in this way are shown in Table 4-5.

More recent treatments of resonance effects modify the simplest bond additivity scheme by allowing different bond properties to be assigned, depending on the hybridizations of the atoms in the bonds [6]. For example, the bond energy of a vinyl C-H is taken as different from that of a methylene C-H. This is the same as the modification to atomic additivity of bond lengths that we discussed above.

More elaborate additivity schemes, involving group additivity, have recently been devised [2] and are remarkably successful. Group values of C_p°, S°, ΔH_f°, and ΔG_f° are assigned to groups such as $-CH_3$, $-CH_2-$, $=CH_2$, etc., and corrections for gauche conformations and cis-interactions of olefins are tabulated. For hydrocarbons and compounds with single functionality, the group additivity scheme allows calculation of all of the thermodynamic properties nearly to within experimental error limits. The basic conclusion to be drawn from the success of this scheme is that the properties of many molecules are determined by nearest neighbor interactions.

TABLE 4-4

Heats of Hydrogenation [4]

Compound	Heat of hydrogenation (kcal/mole of H_2)
$H_2C=CH_2$	32.8
$CH_3CH=CH_2$	30.1
cis-2-butene	28.6
trans-2-butene	27.6
$(CH_3)_2C=CH_2$	28.4
$(CH_3)_2C=C(CH_3)_2$	26.6

TABLE 4-5

Resonance Energy Calculated From Heats of
Hydrogenation Compared to Isolated Double Bonds [5]

Compound	Resonance energy (kcal/mole)
Benzene	48.6
Styrene	51.0
Azulene	46.0
Cyclooctatetraene	4.8
Pyridine	37.0
Phenol	50.0
Aniline	51.0
Benzaldehyde	47.0
Propionic acid	24.0
p-Benzoquinone	16.0

It is worth noting that all of the additivity schemes which we have considered predict no change between free energies of reactants and products for a reaction involving an exchange of groups where nearest neighbor interactions remain unchanged on going from reactants to products [Eq. (4-1)].

$$A\text{-}G\text{-}X + B\text{-}G\text{-}Y \rightleftharpoons A\text{-}G\text{-}Y + B\text{-}G\text{-}X \qquad (4\text{-}1)$$

That is, even group additivity of free energies predicts that the equilibrium constant for such an exchange reaction should be unity. If A, B, X, and Y are polar groups, or if G is a group whose conformation brings A, B, X, and Y into close proximity, we have already stated that group additivity fails. We shall see that such cases are exactly the ones in which we are most interested when we discuss substituent effects on organic reactions.

4-3. THE PRINCIPLE OF ELECTRONEGATIVITY

A different approach to the problem of estimating molecular properties, and one more pertinent to the case of exchange reactions, is represented

by the concept of electronegativity [7]. In its simplest form, electronegativity is just a correction to additivity of atomic properties for bond
energies. The original definition of electronegativity,

$$(\chi_X - \chi_Y)^2 = D_{X-Y} - \frac{D_{X-X} + D_{Y-Y}}{2} \tag{4-2}$$

where D_{X-X}, D_{Y-Y}, and D_{X-Y} are bond dissociation energies expressed
in electron volts, postulated that the difference between the bond energy of
an X-Y bond and the average of the X-X and Y-Y bonds depends on the
difference in some atomic properties of the atoms, χ_X and χ_Y, called
electronegativity. It was found empirically that the simplest function which
could be used involved the square of the difference in electronegativities.
We may note, in fact, that the square function is the simplest function which
gives a nonzero value of $\Delta H°$ for an exchange reaction:

$$A-B + X-Y \rightleftharpoons A-X + B-Y \tag{4-3}$$

Setting

$$-\Delta H° \text{ (kcal/mole)} = 23.06(D_{AX} + D_{BY} - D_{AB} - D_{XY})$$

a little algebra with Eq. (4-2) shows that

$$\Delta H° = -2 \cdot 23.06 (\chi_X - \chi_B)(\chi_Y - \chi_A) \tag{4-4}$$

Since atomic additivity applies quite well to $S°$, at least in the gas phase,
$\Delta S°$ for the exchange reaction should be close to 0, and, since then $\Delta H° = \Delta G°$,

$$\ln K_{ex} = -\frac{46.1}{RT} (\chi_X - \chi_B)(\chi_A - \chi_Y) \tag{4-5}$$

we shall see in later discussion that this equation has the same form as the
very useful Hammett and Taft equations for substituent effects on reactivity.
   The straightforward application of electronegativity does not help in the
case of the exchange reaction (4-1). Since A, B, X, and Y remain bonded
to the same atoms in both reactants and products, $\Delta H°$ for reaction (4-1)
is calculated to be 0 by application of Eq. (4-2).
   More recent treatments of electronegativity consider the hybridization of
atoms, allowing different electronegativity values for sp^3, sp^2, and sp
hybridized carbon [8], for example. Other modifications allow changes in
electronegativity values depending on the electronegativities of other atoms
bonded to the atom in question [8,9]. For example, the electronegativity
assigned to the carbon in consideration of the C-X bond of $F-CH_2-X$ would

be somewhat greater than that of the carbon in H-CH$_2$-X. These modifications allow application of the concept to reaction (4-1), and make the concept of electronegativity indistinguishable from the Taft equation which we shall consider below.

4-4. THE HAMMETT EQUATION [10,11]

Regularities of substituent effects on the reactivities of organic molecules were first observed in the reactions of m- and p-substituted benzene derivatives. For example, for virtually all reactions of benzene derivatives, it was observed that if the introduction of a p-methyl group increased the rate or equilibrium constant for a reaction, the introduction of a p-cyano group decreased the same rate or equilibrium constant, and that a p-nitro group caused an even larger decrease. The quantitative nature of this regularity is exhibited by the observation that plots of the log of the rate or equilibrium constant for one reaction vs the log of the rate or equilibrium constant for another reaction define quite good straight lines. For example, if the log of the equilibrium constants for the ionizations of a series of m- and p-substituted benzoic acids in water are plotted against the log of the rate or equilibrium constants for the reactions of a series of m- and p-substituted benzenediazonium ions with benzenesulfinate ion in methanol solution, linear relationships are observed. The data are shown in Table 4-6 and the plots are shown in Figure 4-1. Analogous plots are obtained for nearly all reactions of m- and p-substituted benzene derivatives.

These linear log-log plots can be expressed by the linear equation

$$\log K_X - \log K_H = \rho (\log K^\circ_X - \log K^\circ_H) \tag{4-6}$$

where K_X and K_H are the equilibrium or rate constants for the X- and H-substituted benzene derivatives, respectively, in some reaction under consideration, the K°'s are the analogous equilibrium constants for the ionization of benzoic acids, and ρ is the slope of the line.

In order to formalize the treatment, we define a parameter σ_X characteristic of the substitutent X:

$$\sigma_X = \log K^\circ_X - \log K^\circ_H \tag{4-7}$$

and rewrite Eq. (4-6)

$$\log (K_X/K_H) = \sigma_X \rho \tag{4-8}$$

This is called the Hammett equation.

TABLE 4-6 [12,13]

$$XC_6H_4COOH \xrightleftharpoons[\text{water, 25°C}]{K_a} XC_6H_4COO^- + H^+$$

$$XC_6H_4N_2^+ + C_6H_5SO_2^- \xrightleftharpoons[K_2]{k_f} XC_6H_4N{=}N{-}\overset{\displaystyle O}{\underset{\displaystyle O}{\overset{\textstyle ||}{\underset{\textstyle ||}{S}}}}C_6H_5$$

MeOH, 25°C

X	$\log K_a$	$\log K_2$	$\log k_f$
H	-4.203	5.16	2.51
p-CH$_3$	-4.373	4.64	2.20
p-Cl	-3.976	6.04	2.99
m-CF$_3$	-3.77	7.14	3.67
m-Cl	-3.830	6.89	3.56
p-CN	-3.54	7.68	4.17
p-NO$_2$	-3.425	8.01	4.28

The definitions in Eqs. (4-6) and (4-7) have set the σ value for hydrogen at 0, and the ρ value for the ionization of benzoic acids at 1.00. Substituents with positive σ values are those which increase the ionization of benzoic acid; those with negative σ values decrease the ionization of benzoic acid. The value of ρ depends on the type of reaction considered; those reactions which are aided by electron-withdrawing substituents (i.e., those with positive σ values) have positive ρ values; those which are aided by electron-donating substituents (i.e., those with negative σ values) have negative ρ values.

It is worth noting that the equilibrium or rate constant ratio in Eq. (4-8) is actually an equilibrium constant for an exchange reaction of the type written in Eq. (4-1) above. For example, consider the ionizations of phenylacetic acids:

$$p\text{-}CH_3C_6H_4CH_2COOH \xrightleftharpoons{K_{p\text{-}Me}} p\text{-}CH_3C_6H_4CH_2COO^- + H^+$$

$$p\text{-}NCC_6H_4CH_2COOH \xrightleftharpoons{K_{p\text{-}CN}} p\text{-}NCC_6H_4CH_2COO^- + H^+$$

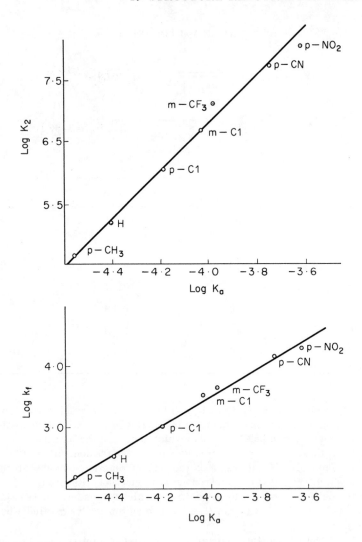

Figure 4-1. Log-log plots of data in Table 6. (a) Equilibrium vs. equilibrium plot. (b) Rate vs. equilibrium plot.

Application of Eq. (4-8) gives

$$\log (K_{p\text{-}CN}/K_{p\text{-}CH_3}) = (\sigma_{p\text{-}CN} - \sigma_{p\text{-}CH_3})\rho \qquad\qquad (4\text{-}9)$$

The ratio $K_{p\text{-}CN}/K_{p\text{-}CH_3}$, however, is just the equilibrium constant for the exchange reaction:

$$p\text{-}NCC_6H_4CH_2COOH + p\text{-}CH_3C_6H_4CH_2COO^- \rightleftharpoons$$

$$p\text{-}NCC_6H_4CH_2COO^- + p\text{-}CH_3C_6H_4CH_2COOH$$

The success of the Hammett equation can undoubtedly be attributed largely to the fact that the reaction sites are separated from m- and p-substituents in a manner that prevents direct interaction of the substituent and reaction site. When such direct interactions are possible, in fact, we find substantial deviations from the behavior predicted by the Hammett equation. Such direct interactions can occur in p-substituted benzenes where one of the substituents has an orbital containing a lone pair of electrons and the other has an empty orbital, both capable of overlap with the benzene π-orbitals. As you have probably already learned, a molecule such as p-nitroaniline has a dipole moment quite different from that expected from the group dipole moments, and shows an absorption spectrum unrelated to that of aniline or nitrobenzene. These and other properties of p-nitroaniline are understandable on the basis of resonance structures:

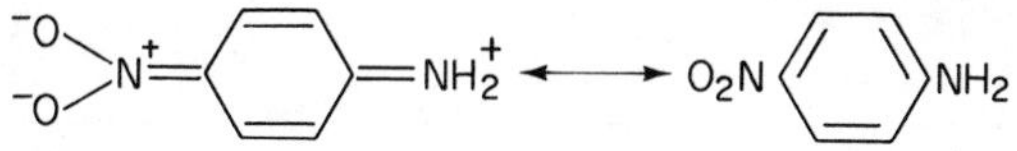

Such direct resonance interactions show up very clearly in the reactions of substituted anilines and substituted phenoxides. The equilibrium constants for ionizations of anilines and phenols, for example, are shown in Table 4-7, and a plot of the pK's of benzoic acids vs the pK's of phenols is shown in Figure 4-2. An excellent straight line plot, in accord with Eq. (4-8), is found for all substituents except those p-substituents for which direct resonance interaction forms can be drawn for the phenoxide ions:

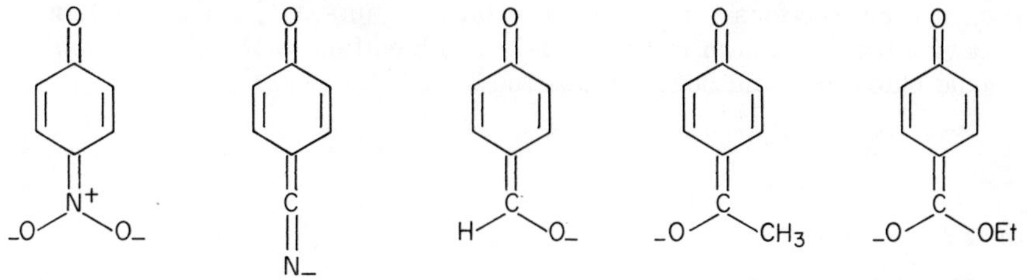

TABLE 4-7

Ionizations of Benzoic Acids, Phenols, and Anilinium Ions[a] [12]

Substituent	pK_{ArCOOH}	pK_{ArOH}	$pK_{ArNH_3^+}$
H	4.203	9.90	4.58
m-OCH$_3$	4.08	9.63	4.21
p-OCH$_3$	4.47	10.40	5.31
m-Br	3.81	9.01	3.51
m-Cl	3.83	9.09	3.50
m-F	3.86	9.25	3.56
m-NO$_2$	3.49	8.32	2.44
m-CN	3.64	8.52	–
p-Br	3.97	9.30	3.86
p-NO$_2$	3.42	7.10	0.93
p-CN	3.54	7.94	–
p-CHO	3.98	7.64	–
p-COCH$_3$	3.80	8.16	–
p-CO$_2$C$_2$H$_5$	3.75	8.54	–

[a] Values are for H_2O solutions at 25°C.

Notice that all meta-substituents, and those para-substituents such as p-CH$_3$O and p-Br which have no unsaturation of the atom bonded to the benzene ring, closely follow Eq. (4-8).

Other reactions of phenoxides and anilines show similar deviations from the Hammett equation, and the deviations are regular enough that we may assign special substituent constants to those groups capable of direct resonance interactions. These substituent constants, σ^-, are defined by drawing the line shown in Figure 4-2, and then finding the σ^- value which would place the solid points on the line.

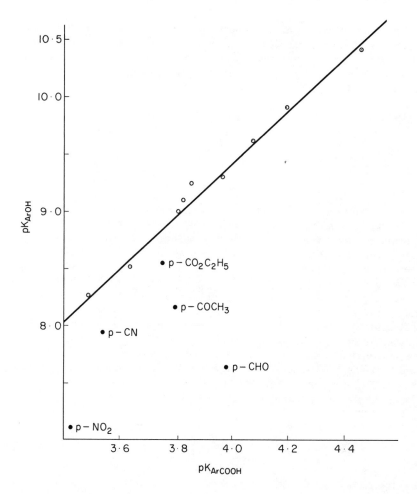

Figure 4-2. Plot of data in Table 4-7.

The same type of direct resonance interaction of a substituent with a reaction site, but in the opposite sense, shows up in the reactions of carbonium ions. For triarylmethyl cations, for example, substituents such as p-CH$_3$O, p-NH$_2$, and p-Cl allow resonance forms such as the following to be drawn:

Another special set of substituent constants, σ^+, are defined for use in reactions where the reaction site group is capable of this direct resonance electron withdrawal from substituents in the para-position which have lone pairs of electrons. The σ^+ values are obtained from the rates of solvolysis of a series of substituted t-cumylchlorides [14]

by a procedure analogous to that described for evaluation of the σ^- constants.

Both σ^- and σ^+ apply only to para-substituents: the σ^- values to those capable of direct resonance electron withdrawal from the reaction site, and the σ^+ values to those capable of direct resonance electron donation to the reaction site. A list of the various substituent constants for some common substituents is given in Table 4-8.

Quite generally, the ordinary σ values are determined by two conceptually different types of effect. The polar effect of a substituent is most conveniently viewed as a group electronegativity, which we mentioned above, or as the result of the dipole moment of the group. In the benzene series, virtually all substituents are also capable of a resonance effect, which arises from resonance interaction of the substituent and the benzene ring. This resonance effect can be either an electron-withdrawing or donating effect, depending on the nature of the substituent:

electron withdrawing   electron donating

Those substituents which require σ^+ values in the reactions of carbonium ions generally have electron-donating resonance forms of the sort written above even in "normal" reactions; those which require σ^- values in reactions of anilines generally have the analogous electron withdrawing structures for "normal" reactions. Note that we are considering here only the interaction of the substituent with the aromatic ring, not the direct resonance with reaction site which requires the use of the σ^+ and σ^- scales, and that this resonance effect is included in the normal σ values. Since these resonance forms place charge in the ortho and para positions, we may expect that the resonance effects contribute more to the para-substituent constants than to the meta-substituent constants.

4-5. THE TAFT EQUATION

In order to attempt a separation of the resonance and polar effects of substituents in benzene reactivities, one would like to have a series of compounds in which the geometrical relationships of reaction site and substituent have the rigidity of the benzene derivatives, and approximately the same spatial relationship, but in which the resonance effect of the substituents cannot operate. The 4-substituted bicyclo[2.2.2]octane-1-carboxylic acids [15]

furnish such a series. The effects of the substituents on the ionizations of these acids were used to define a scale of substituent parameters, σ_I, in a manner analogous to the definition of σ from benzoic acid ionizations:

$$\sigma_I = \log (K_x/K_H) \tag{4-10}$$

It was then shown that these σ_I parameters could be used in an equation analogous to the Hammett equation

$$\log (k_x/k_H) = \sigma_I \rho_I \tag{4-11}$$

to correlate rates of reactions of the acids with diphenyldiazomethane and rates of reactions of the corresponding esters in base-catalyzed hydrolysis.

As we might expect from the above discussion of resonance effects, the σ_I values are not proportional to the Hammett σ values for either meta or para substituents. It is found, in accordance with expectation, that the relationship

$$\sigma_{para} - \sigma_I = 3(\sigma_{meta} - \sigma_I) \tag{4-12}$$

holds approximately for various substituents. Other work has shown that these differences between Hammett σ values and the σ_I values provide a reasonable measure of the resonance interaction of the substituents with the benzene ring.

The σ_I values of substituents are believed to be measures of the polar effect of substituents in all aliphatic reactivities. For both practical and historical reasons, however, most of the σ_I values now available did not result from studies of the bicyclo[2.2.2]octane derivatives. Practically, the necessary derivatives are too difficult to synthesize. Historically, it turns out that nature, at least in this instance, was cooperative.

Since acyclic aliphatic derivatives generally have a number of possible conformations which can bring parts of a chain separated by several methylene groups into close proximity, it was suspected that substituent effects on reactions of such compounds would have widely variable steric and polar contributions. By a study of both acid- and base-catalyzed hydrolysis of substituted acetate esters, it was hoped that this difficulty could be overcome, allowing an evaluation of polar effects of the substituents [16]. The mechanisms of acid- and base-catalyzed ester hydrolysis had already been fairly well established:

$$XCH_2C\overset{\diagup O}{\underset{\diagdown OEt}{}} + H^+ \longrightarrow \left[XCH_2 - \overset{\overset{\displaystyle OH}{|}}{\underset{\underset{\displaystyle OH_2^+}{|}}{C}} - OEt \right]^{\neq}$$

$$XCH_2C\overset{\diagup O}{\underset{\diagdown OEt}{}} + OH^- \longrightarrow \left[XCH_2 - \overset{\overset{\displaystyle O^-}{|}}{\underset{\underset{\displaystyle OH}{|}}{C}} - OEt \right]^{\neq}$$

The transition states for the two paths are quite similar except for charge and the number of protons on the oxygens. In particular, the conversion of the ester into the transition states should involve the same steric effects for both acid and base catalysis. Because of the difference in charge of the transition states, however, the polar effect of substituents should operate in different directions for the two reactions. It was therefore hoped that the quantity

$$\log (k_{XCH_2}/k_{CH_3})_{OH} - \log (k_{XCH_2}/k_{CH_3})_{H^+}$$

where the first ratio of rate constants is for base catalysis and the second is for acid catalysis, would give a measure of the polar effect of the substituent X relative to hydrogen. This quantity turns out, in fact, to be proportional to the σ_I values determined from the bicyclo[2.2.2]octane series. Since many more substituents are available in substituted acetates than in the bicyclic compounds, an alternative definition [17] of σ_I [Eq. (4-13)] has been used to determine most of the values reported in Table 4-8.

TABLE 4-8

Substituent Parameters [17]

Substituent	σ_I	σ_{meta}	σ_{para}	σ^+	σ^-
$-CH_2CN$	0.23	–	0.01	–	–
$-CH_2Cl$	0.17	–	0.18	-0.01	–
$-CH_2CH_3$	0.00	-0.07	-0.15	-0.30	–
$-CH_3$	0.00	-0.07	-0.17	-0.31	–
$-CF_3$	0.41	0.43	0.54	–	–
$-CHO$	0.31	0.35	0.22	–	1.13
$-COCH_3$	0.28	0.38	0.50	–	0.87
$-CONH_2$	0.21	0.28	–	–	0.62
$-COOR$	0.30	0.37	0.45	–	0.68
$-C_6H_5$	0.10	0.04	0.00	-0.17	–
$-COO^-$	-0.14	-0.10	0.00	-0.03	–
$-COOH$	0.34	0.37	0.45	–	–
$-CN$	0.56	0.56	0.66	–	0.90
$-N(CH_3)_3^+$	0.92	0.88	0.82	0.41	–
$-NO_2$	0.63	0.71	0.78	–	1.24
$-N(CH_3)_2$	0.10	-0.21	-0.83	-1.7	–
$-NH_2$	0.10	-0.16	-0.66	-1.3	–
$-N_3$	0.44	0.33	0.08	–	0.11
$-OCH_3$	0.25	0.12	-0.27	-0.78	–
$-OH$	0.25	0.12	-0.37	-0.92	–
$-F$	0.52	0.34	0.06	-0.07	-0.02
$-Cl$	0.47	0.37	0.23	0.11	–
$-Br$	0.45	0.39	0.23	0.15	–

$$\sigma_I = 0.182 \, [\text{Log}(k_{XCH_2}/k_{CH_3})OH - \text{Log} \, (k_{XCH_2}/k_{CH_3})H^+] \qquad (4\text{-}13)$$

The σ_I values correlate the reactions of many aliphatic derivatives, such as the ionizations of α-substituted acetic acids, the ionizations of aliphatic amines, and the ionizations of aliphatic alcohols, by the use of Eq. (4-11), which is commonly called the Taft equation.

4-6. INTERPRETATION OF ρ VALUES

Thus far we have discussed σ values, but have only stated that ρ's are slopes of the linear correlations. In actual fact, the ρ values provide valuable information about reaction mechanisms, and it is just as important to understand these as the σ's. Table 4-9 contains a list of some representative ρ values.

Qualitatively we may state that reactions which are aided by electron-withdrawing substituents, such as the ionization of acids, will have positive ρ values, and vice versa. Several comparisons in Table 4-9, for example the ρ values for benzoic, phenylacetic, and phenylpropionic acids, show that the ρ value decreases by approximately a factor of two for each methylene group inserted between the substituent and the reaction site. We also note that ρ values are solvent dependent, and appear to depend more on hydrogen bonding than on dielectric constant for the ionizations of benzoic acids, since MeOH, DMSO, and CH$_3$CN have nearly the same dielectric constants. We may further note that the decrease in ρ by a factor of approximately two for each methylene group also seems to hold fairly well when we compare the ionization of benzoic acids, anilines, and pyridinium ions. That is, the factor of two seems to apply as we move the charge closer to the substituent, regardless of what the separating atoms are. Thus, the ρ value depends on the location of charge produced or neutralized on going from reactant to product, and by implication (which can be verified by examining the ionizations of arylphosphoric acids) the ρ value depends on the magnitude of the charge.

In the correlations of rates of reactions, the two states pertinent to the ρ values are the reactant and transition states. Reversing the above arguments, then, the magnitude of ρ for a reaction gives us information on the location and magnitude of charge at the transition state. Before returning to a discussion of the mechanisms of solvolysis reactions, we may perform a little algebra to strengthen the basis for the interpretation of ρ values. We have stressed in Eq. (4-9) that the ρ-σ relationships actually apply to the calculation of the equilibrium constant for the exchange reaction (4-1):

$$A\text{-}G\text{-}X + B\text{-}G\text{-}Y \xrightleftharpoons{K_{ex}} A\text{-}G\text{-}Y + B\text{-}G\text{-}X$$

TABLE 4-9

ρ Values

Reaction	ρ
XCH_2COOH ionization, H_2O, 25°C, equil. [16]	$\rho_I = 3.8$
$XCH_2CH(OEt)_2$ acid-catalyzed hydrolysis, 50% aqueous dioxane, 25°C, rate [16]	$\rho_I = 8.1$
XCH_2CH_2OH ionization, 2-propanol, 27°C, equil. [16]	$\rho_I = 3.0$
$XCH_2CH\text{-}OTs$ solvolysis, HOAc, 30°C, rate [16,18] $\quad \underset{CH_3}{\mid}$	$\rho_I = -7.7$
$\overset{CH_3}{\underset{CH_3}{\mid \; \mid}}$ $XCH_2C\text{-}Cl$ solvolysis, 80% aqueous EtOH, 25°C, rate [18]	$\rho_I = -7.3$
XCH_2CH_2OTs solvolysis, EtOH, 100°C, rate [18]	$\rho_I = -1.6$
$XCH_2CH_2Br + C_6H_5S^-$, MeOH, 20°C, rate [18]	$\rho_I = -1.4$
X—(bicyclooctane)—$N - H^+$ ionization, H_2O, 25°C, equil. [19]	$\rho_I = 5.1$
X—(bicyclooctane)—$COOH$ ionization, 50% aqueous EtOH, 25°C, equil. [15]	$\rho_I = 1.00$
$ArCH_2CH_2COOH$ ionization, H_2O, 25°C, equil. [10]	$\rho = 0.24$
$ArCH_2COOH$ ionization, H_2O, 25°C, equil. [10]	$\rho = 0.56$
$ArCOOH$ ionization, H_2O, 25°C, equil. [10]	$\rho = 1.00$
$ArCOOH$ ionization, MeOH, 25°C, equil. [10]	$\rho = 1.54$
$ArCOOH$ ionization, EtOH, 25°C, equil. [10]	$\rho = 1.65$
$ArCOOH$ ionization, DMSO, 25°C, equil. [20]	$\rho = 2.6$
$ArCOOH$ ionization, CH_3CN, 25°C, equil. [20]	$\rho = 2.8$
3-, and 4-X pyridineH$^+$, ionization, H_2O, 25°C, equil. [19]	$\rho = 5.2$
$ArN(CH_3)_2 + CH_3I$, 90% acetone, 35°C, rate [10]	$\rho = -3.3$
$ArOH$ ionization, H_2O, 25°C, equil. [10]	$\rho = 2.1$
$ArNH_3^+$ ionization, H_2O, 25°C, equil. [10]	$\rho = 2.8$
$ArCOCl$ hydrolysis, 95% acetone, 25°C, rate [10]	$\rho = 1.8$

TABLE 4-9 (cont.)

Reaction	ρ
t-Cumylchloride solvolysis, 90% acetone, 25°C, rate [14]	$\rho = -4.5$
t-Cumylchloride solvolysis, EtOH, 25°C, rate [14]	$\rho = -4.7$
Triarylmethanol + H^+, H_2O, 25°C, equil. [4]	$\rho = -3.6$
$ArSO_3Me$ solvolysis, EtOH, 70°C, rate [21]	$\rho = 1.3$
$ArSO_3Et$ solvolysis, EtOH, 70°C, rate [21]	$\rho = 1.3$
$ArSO_3CH(CH_3)_2$ solvolysis, EtOH, 70°C, rate [21]	$\rho = 1.6$
$ArSO_3(2\text{-Ad})^a$ solvolysis, EtOH, 70°C, rate [21]	$\rho = 1.8$
$ArSO_3$ (cyclohexyl) solvolysis, HOAc, 70°C, rate [21]	$\rho = 1.3$
$ArSO_3$ (cyclobutyl) solvolysis, HOAc, 25°C, rate [21]	$\rho = 1.4$

[a] 2-adamantyl arenesulfonates

If we are dealing with rates, Y is the reaction site group at the transition state for the reaction, and X is the reaction site group in the reactant state.

Considering A and B as substituents and X and Y as reaction site groups, the application of either Eq. (4-8) or (4-11) gives

$$\log K_{ex} = (\sigma_A - \sigma_B)\rho_{XY} \qquad (4\text{-}14)$$

If there is generality to the Hammett or Taft relationships - in particular, if it is possible to assign σ values to all groups - we may also view the exchange reaction as one in which X and Y are substituents and A and B are the reaction site groups. Application of Eq. (4-8) or (4-11) then gives

$$\log K_{ex} = (\sigma_X - \sigma_Y)\rho_{AB} \qquad (4\text{-}15)$$

Therefore,

$$(\sigma_A - \sigma_B)\rho_{XY} = (\sigma_X - \sigma_Y)\rho_{AB} \qquad (4\text{-}16)$$

which, on rearrangement, gives

$$\frac{\sigma_A - \sigma_B}{\rho_{AB}} = \frac{\sigma_X - \sigma_Y}{\rho_{XY}} \qquad (4\text{-}17)$$

Now, since A and B can be varied independently of X and Y, Eq. (4-17) implies

$$\rho_{XY} = \beta_G (\sigma_X - \sigma_Y) \qquad \qquad (4\text{-}18)$$

where β_G is a constant characteristic of the group G. This means that the ρ value for a reaction is proportional to the change in σ value of the reaction site group on going from reactants to products in an equilibrium, or from reactants to transition state in a rate process.

We can evaluate β_G for the meta-substituted benzene ring from the fact that ρ for the ionization of benzoic acids in water is 1.00, and from the σ_m values of the COOH and COO$^-$ groups listed in Table 4-8. The value obtained from Eq. (4-18) is $\beta_{m\text{-}C_6H_4} = 2.1$. Similarly, for a CH_2 group, we use the ρ_I for the ionization of acetic acids and the σ_I values in Table 4-8 to obtain $\beta_{CH_2} = 7.9$. These values along with the σ values in Table 4-8 allow one to calculate ρ values for a very large number of reactions. We can also calculate the fall-off in ρ value on the insertion of a methylene group by looking at the fall-off in σ values on going from X to X-CH_2. The approximate fall-off factor found in this manner is 2.5 per interposed methylene group.

Because of the solvent dependence of the ρ and σ values of charged groups, it is difficult to carry out accurately self-consistent calculations of the sort outlined in the last paragraph. The estimation or semiquantitative interpretation of ρ values, however, rests on a firm foundation.

We may now examine the solvolysis reactions listed in Table 4-9 to see if the ρ values are consistent with the mechanisms of these reactions which we have discussed in previous sections.

We may first note the ρ value for the solvolysis of t-cumylchlorides in 90% acetone is -4.5. The reaction of triarylmethanols with acid is an equilibrium process forming the triarylmethyl cations as products, and this reaction in water has $\rho = -3.6$. Even allowing a substantial increase in magnitude of ρ on going from water to 90% acetone, the value of ρ for the solvolysis reaction must be interpreted as indicating virtually a full unit positive charge on the benzylic carbon at the transition state for this reaction. We may reach the same conclusion by examining the behavior of the aliphatic compounds which we have classified as reacting by the S_N1 pathway. The ionization of alcohols has $\rho_I = 3.0$ in isopropanol solvent. Since this is a reaction which places the charge three atoms from the substituent, we may expect that a reaction $X\text{-}CH_2\text{-}C \rightarrow X\text{-}CH_2\text{-}C^+$ should have $\rho_I = -7.5$ (i.e., -2.5 x 3.0). The solvolysis of tertiary chlorides in 80% ethanol has $\rho_I = -7.3$.

The solvolysis of 2-propyl tosylates in acetic acid has $\rho_I = -7.7$, clearly indicating that this is an S_N1 type of reaction for this secondary substrate. The ρ_I value for solvolysis of the ethyl tosylates in ethanol is considerably smaller in magnitude, only -1.6, showing the expected clear distinction of the primary substrates from the tertiary. The negative value of ρ_I indicates that there is a small amount of positive charge (naively, we can estimate $1.6/7.5 = 0.2$ units) at the substituted carbon at the transition state for this S_N2 reaction. The positive character of the carbon is in accordance with expectations based on viewing the transition state as a pentavalent carbon species, since such hypervalent compounds as sulfuranes appear to have substantial positive charge on the hypervalent atom.

Although it appears that the ρ values for the above reactions are consistent with our proposed mechanisms, and provide some refinement of the picture of the transition states, there are some other data in Table 4-9 more difficult to rationalize. The last six entries in the table show the effect of substitution in the arenesulfonate moiety on the solvolysis rates of various alkyl arenesulfonates. The ρ values should reflect primarily the amount of negative charge developed on the leaving sulfonate anion. The rather surprising fact is that these ρ values vary only slightly for different alkyl groups. It is not at all obvious how this information can be made consistent with the mechanisms which we have discussed. It is particularly bothersome that the ρ value for 2-adamantyl arenesulfonates, which must solvolyze by an S_N1 mechanism, is only slightly greater than the ρ value for methyl and ethyl arenesulfonates, which must be classified as S_N2 substrates. The data may be indicative of solvent interactions with the leaving group, but such rationalizations must be strictly ad hoc until more information becomes available.

4-7. STERIC EFFECTS

In view of the observed broad generality of Eqs. (4-8) and (4-11) for treating polar effects of substituents, one might hope that similarly broad relationships could be found for treating steric effects on reactivity. Although considerable progress has been made in understanding steric effects, no treatment having the precision or generality comparable to those for polar effects yet exists. Part of the problem is that steric effects have a number of different sources and modes of operation.

The data in Table 4-10 can serve as a basis for illustrating several effects of steric origin. The behaviors of the very small cyclopropyl and cyclobutyl compounds can be largely attributed to ring angle strain. The heats of combustion drop quite regularly from cyclopropane to cyclohexane as the ring angle expands from 60° to the normal tetrahedral angle. The interpretation of the solvolysis rates of the cycloalkyl tosylates cannot be made on the basis of angle strain alone. The conversion of the reactant to the carbonium-ion-like transition state must involve an increase in angle strain for the small rings because the preferred angle of 120° for the car-

TABLE 4-10

Some Properties of Cyclic Compounds: Heats of Combustion of
Cycloalkanes and Rates of Acetolysis of Cycloalkyl Tosylates [22]

Ring size	ΔH°_{comb} per CH_2 (kcal)	$k_{acetolysis}$ relative to cyclohexyl tosylate (60 °C)
3	166.6	2×10^{-5}
4	164.0	14
5	158.7	16
6	157.4	1.00
7	158.3	31
8	158.6	285
9	158.8	266
10	158.6	539
11	158.4	67
12	157.7	–
14	157.4	–

bonium ion is even larger than the tetrahedral angle preferred by the reactant. This probably accounts for the extremely slow reaction of the
cyclopropyl tosylate, but, obviously, other factors must come into play for
the cyclobutyl, cyclopentyl, and cyclohexyl compounds.

The conversion of ketones into alcohols by reduction with sodium boro-
hydride might be expected to show the same angle strain effects, but in a
reverse direction, as the conversion of tosylates into carbonium ions,
since the reduction converts a trigonal carbon into a tetrahedral one [23].
The rate of reduction of cyclobutanone is about forty times faster than that
of cyclopentanone, in accord with the angle strain hypothesis.

We may note that the heat of combustion of cyclohexane shows that this
ring has less total "strain" than any of the cyclic C_5 through C_{12} alkanes.
Careful examination of models indicates that the stability of the C_6 system
probably arises from the fact that the cyclohexane chair conformation gives
excellent accommodation of both the tetrahedral angle and of perfectly
staggered conformations of each of the methylene groups:

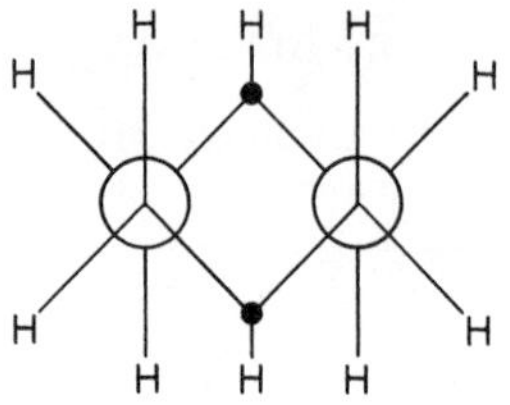

Neither the C_5 nor any of the C_7-C_{11} cycloalkanes can satisfy both of these needs simultaneously. The unfavorable eclipsing of neighboring C-H bonds is most likely the other factor, in addition to the ring angle strain, which influences the solvolysis rates. That is, at least part of this eclipsing is relieved on going from the reactant to the transition state for these compounds.

Another type of steric interaction becomes apparent in cyclic compounds when we consider substitution at one of the ring hydrogens. In the cyclohexyl system, for example, a substituent X can occupy either an axial position or an equatorial position:

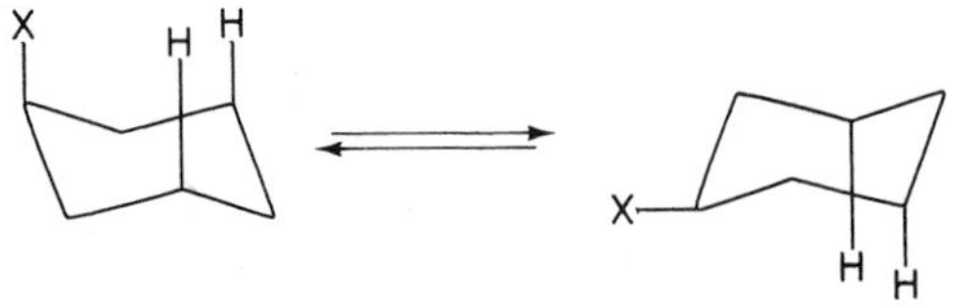

In either position, the C-X bond is staggered with respect to the neighboring C-H bonds. In the axial position, however, the group X comes into close proximity to the hydrogens at C_3 and C_5. By the use of NMR spectroscopy, it is possible to measure the ratio of the two conformers and thereby obtain the free energy change for the conversion, which is a measure of the steric size of the group X. Some such $\Delta G°$ values for the axial to equatorial conversion are shown in Table 4-11.

The constancy of $\Delta G°$ for the halogens is quite striking in view of the known large variations in covalent radii, and illustrates another complication in understanding steric effects. The probable explanation of this behavior is that as the size of the halogen increases, so does the C-X bond length, thus moving the center of the X group farther away from the 3 and 5 hydrogens, exactly counterbalancing the size effect.

The steric size effect measured by $\Delta G°$ can also be important in other ring systems where conformations of the ring may force transannular interactions very similar to the 1,3 interactions in the cyclohexane derivatives.

TABLE 4-11

$\Delta G°$ for Conversion of Axial to Equatorial Cyclohexyl-X [22]

X	$-\Delta G°$ (kcal)	X	$-\Delta G°$ (kcal)
F	0.2	OTs	0.7
C≡CH	0.2	OCH_3	0.7
CN	0.2	SH	0.9
Cl	0.4	COOR	1.1
Br	0.4	NH_2	1.4
I	0.4	CH_3	1.7
OH	0.8	CH_2CH_3	1.8
OAc	0.7	$i-C_3H_7$	2.1

A somewhat different steric size effect is exhibited in S_N2 type reactions. In these cases the conversion of reactant into transition state may be viewed as the conversion of a four-coordinated carbon into a five-coordinated carbon. Thus, larger groups attached to the reacting carbon will cause more crowding in the transition state for the reaction. The relative rate constants for the reactions of ethyl, n-propyl, i-butyl, and neopentyl bromides with iodide ion in acetone are 100, 82, 3.6, and 1.2×10^{-3}, respectively, clearly showing this crowding influence on the S_N2 reaction.

The effect of steric bulk on the rates of S_N1 reactions is expected to be in the opposite direction from that on the S_N2 reactions, since the S_N1 reaction involves conversion of a four-coordinated to a three-coordinated carbon, thereby relieving crowding. In fact, the solvolysis of tri-t-butyl-methyl p-nitrobenzoate is an extremely fast reaction in comparison to other tertiary systems [24].

4-8. DEUTERIUM AS A SUBSTITUENT

The study of reaction mechanisms by variation of substituents to probe polar and steric effects is one of the most commonly used and frequently tested techniques of physical organic chemistry. Problems sometimes arise, however, in which it is conceivable that a substantial change in either steric or polar effects of substituents will change the mechanism of the reaction. We have already discussed one such case, the solvolysis of the 3-aryl-2-butyl brosylates, in which substituents completely alter the stereochemistry and product distribution of the reaction. The reactions of benzyl halides and tosylates, in which rates of solvolysis are increased by both electron-withdrawing and electron-donating substituents, is quite possibly another such example.

The mechanisms of the solvolyses of secondary systems, as we have already seen in earlier sections, show characteristics of both S_N1 and S_N2 pathways, and we may suspect that such systems would be particularly disposed to change mechanism if a large change in structure were made. This suspicion is strengthened by the data shown in Table 4-12 for the rates of solvolysis of isopropyl brosylate and pinacolyl brosylate (3,3-dimethyl-2-butyl brosylate).

The simplest interpretation of this data is that the isopropyl brosylate solvolysis is an S_N1 process in the highly ionizing but poorly nucleophilic solvents CF_3COOH and CF_3CH_2OH, and an S_N2 process in the more nucleophilic water and ethanol solvents, while the pinacolyl brosylate solvolysis is an S_N1 process for all solvents. This interpretation is strengthened by similar studies of the relative rates of solvolysis of isopropyl tosylate and 2-adamantyl tosylate in several solvents [26]. The 2-adamantyl tosylate

TABLE 4-12

Relative Rates of Solvolysis of Isopropyl and
Pinacolyl Brosylates [25]

Solvent	k_{i-Pr}/k_{Pin}	$(k_H/k_{\alpha-D})_{i-Pr}^a$
CF_3COOH	3.6×10^{-4}	1.22
97% CF_3CH_2OH[b]	2.6×10^{-2}	1.16
70% CF_3CH_2OH[b]	1.07×10^{-1}	1.14
50% CF_3CH_2OH[b]	1.72×10^{-1}	1.12
50% CH_3CH_2OH[b]	7.14×10^{-1}	1.11
80% CH_3CH_2OH[b]	2.27	1.10
90% CH_3CH_2OH[b]	3.35	1.08

[a] Relative rates of solvolysis of isopropyl brosylate and α-deuterio isopropyl brosylate.

[b] The other component of the solvent is water.

has a structure in which the back side of the substituted carbon is so blocked that we may rule out the possibility of an S_N2 reaction for this compound. The rate constant ratios, k_{i-Pr}/k_{2-Ad}, for solvolysis of the tosylates in various solvents are CF_3COOH, $10^{-2.5}$; HCOOH, $10^{0.5}$; CH_3COOH, $10^{1.1}$; 50% aqueous EtOH, $10^{1.1}$; EtOH, $10^{3.0}$. Thus, it seems quite likely that the substitution of a t-butyl group for a methyl group of isopropyl arenesulfonates in at least some solvents changes the mechanism of solvolysis.

In such cases we would like to have a very delicate probe which can tell us something about the transition state for the particular compound of interest. The second column of Table 4-12 pertains to data obtained with such a probe: the substitution of a deuterium atom for a hydrogen atom at the reacting carbon. Since hydrogen and deuterium have the same electronic structure, this substitution does not affect the potential energy of either the reactant or the transition state. We will discuss isotope effects in considerably greater detail in Chapters 8 and 9, but will simple note here that the difference in mass of the isotopes leads to a difference in vibrational energy of the vibrations in which the isotopes move. The difference in vibrational energy of hydrogen and deuterium depend primarily on the force constant for the particular vibration. Thus, any change in the force constants associated with the C-H bond on going from reactant to transition state will cause the vibrational energy difference between hydrogen and deuterium to change, and we will see an effect on the relative rate constants for the hydrogen- and deuterium-substituted compounds. Empirically, the S_N1 solvolyses of α-deuterio alkyl tosylates and brosylates give $k_H/k_{\alpha-D} = 1.22$, and of alkyl chlorides give $k_H/k_{\alpha-D} = 1.16$. The S_N2 solvolyses usually give $k_H/k_{\alpha-D}$ very close to unity, 0.98 to 1.05, approximately. An example of the use of this technique, and some further examples of changes in mechanism brought about by substitution or by solvent, are shown in Table 4-13.

TABLE 4-13

α-Deuterium Isotope Effects in the Solvolyses of $X-C_6H_4CH_2OBs$ [27]

Solvent[a]	$k_H/k_{\alpha-D}$		
	X=H	X=p-CF_3	X=p-NO_2
97% CF_3CH_2OH	1.17	–	1.03
80% CF_3CH_2OH	1.16	1.04	–
70% CF_3CH_2OH	–	1.04	1.01
70% Ethanol	–	1.02	1.01
80% Ethanol	1.07	1.02	1.00

TABLE 4-13 (cont.)

Solvent[a]	$k_H/k_{\alpha-D}$		
	X=H	X=p-CF$_3$	X=p-NO$_2$
90% Ethanol	1.06	1.01	1.00
95% Ethanol	1.05	1.01	–

[a] The other solvent component is water in all cases.

PROBLEMS

1. In Table 3-6, are listed rate constants for the acetolyses of eight 3-aryl-2-butylbrosylates. Make a Hammett plot of the data. Assuming that the p-NO$_2$ and m-CF$_3$ derivatives have no aryl participation, calculate k_Δ/k_{2s} for the other compounds, where k_Δ and k_{2s} are defined as in Problem 1 of Chapter 3.

2. Estimate the ρ value expected for the acetolysis of 2-aryl-1-methyl-cyclopentyl tosylates. Estimate the ρ value for the ionizations of trans-4-X-cyclohexane carboxylic acids in water. Note the ρ_I values for ionizations of acetic acids and for ionizations of the 4-X-bicyclo[2.2.2] octane 1-carboxylic acids, and make a reasonable interpolation from these for the cyclohexyl case.

3. The pK$_a$ of phenylacetic acid is 4.31 and that of acetic acid is 4.76. From this information and the ρ_I value for ionization of acetic acids, calculate the σ_I value for the phenyl group. Using the ρ value for the ionization of substituted phenylacetic acids and the σ_I value for the phenyl group, derive an expression which relates the σ_I value of a substituted phenyl group to the Hammett σ values of the substituents.

4. On the basis of the ρ value in Table 4-9 and the fact that the reactions are first-order with respect to [H$^+$], propose a mechanism for the hydrolysis of acetals: X-CH$_2$CH(OEt)$_2$.

5. The relative rates of reaction of a series of compounds, X-CH$_2$Br, with iodide ion in acetone are

X	k_{rel}
CH_3	1.00
F	0.79
Cl	0.13
Br	0.041
I	0.059

Assuming that ρ_I is the same as for attack by $C_6H_5S^-$ in methanol, evaluate the steric effects of the various halogens relative to the methyl group.

6. In Table 3-4 the relative rate constants for solvolysis of a series of compounds, $X-CH_2CH_2OTs$, are given. Evaluate the rate enhancements due to neighboring group participation for each substituent.

7. Assuming that the solvolyses are S_N1 reactions, calculate the relative rate constants for the solvolyses of cis- and trans-4-t-butylcyclohexyl tosylates. (ΔG° for the t-butyl group is more negative than −4.)

8. Assuming that ΔS° is constant for a reaction series, derive an expression for the temperature dependence of ρ.

9. What conclusions can you draw from the following data;

$$H_3C \xrightarrow{1.53\text{Å}} CH_3 \qquad H_3C \xrightarrow{1.51\text{Å}} C\overset{H}{\underset{CH_2}{\diagdown}} \qquad H_3C \xrightarrow{1.46\text{Å}} C \equiv CH$$

$$\underset{H}{\overset{H_2C}{\diagdown}}C \xrightarrow{1.47\text{Å}} C\overset{CH_2}{\underset{H}{\diagup}} \qquad \underset{H}{\overset{H_2C}{\diagdown}}C \xrightarrow{1.43\text{Å}} C \equiv CH \qquad HC \equiv C \xrightarrow{1.37\text{Å}} C \equiv CH$$

REFERENCES

1. S. W. Benson and J. H. Buss, J. Chem. Phys., 29, 546 (1958).

2. S. W. Benson, F. R. Cruickshank, D. M. Golden, G. R. Haugen, H. E. O'Neal, A. S. Rodgers, R. Shaw, and R. Walsh, Chem. Revs., 69, 279 (1969).

3. L. N. Ferguson, The Modern Structural Theory of Organic Chemistry, Prentice-Hall, Inc., Englewood Cliffs, New Jersey, 1963, Chaps. 1-2. These chapters contain an unusually thorough discussion of bond parameters and a fuller listing of data.

4. J. Hine, Physical-Organic Chemistry, McGraw-Hill Book Co., Inc., New York, New York, 1962. Contains more complete data.

5. L. N. Ferguson, ibid., pp. 317-319. A slightly different method of calculation is given.

6. M. J. S. Dewar and H. N. Schmeising, Tetrahedron, 11, 96 (1960).

7. L. Pauling, The Nature of the Chemical Bond, 3rd ed., Cornell University Press, Ithca, New York, 1960.

8. P. Wells, Prog. Phys. Org. Chem., 6, 111 (1968).

9. J. Hinze, M. A. Whitehead, and H. H. Jaffe, J. Amer. Chem. Soc., 85, 148 (1963).

10. L. P. Hammett, Physical Organic Chemistry, 1st ed., McGraw-Hill Book Co., Inc., New York, New York, 1940, Chap. 7.

11. L. P. Hammett, Physical Organic Chemistry, 2nd ed., McGraw-Hill Book Co., Inc., New York, New York, 1970, Chap. 11.

12. J. W. Larson and L. G. Hepler, Solvent-Solute Interactions, (J. F. Coetzee and C. D. Ritchie, eds.), Marcel Dekker, Inc., New York, New York, 1969, Chap. 1. A more complete listing of pK's are given.

13. C. D. Ritchie, J. Saltiel, and E. S. Lewis, J. Amer. Chem. Soc., 83, 4601 (1961).

14. H. C. Brown and Y. Okamoto, J. Amer. Chem. Soc., 80, 4979 (1958).

15. J. D. Roberts and W. T. Moreland, J. Amer. Chem. Soc., 75, 2167 (1953).

16. R. W. Taft in Steric Effects in Organic Chemistry, (M. S. Newman, ed.), John Wiley and Sons, Inc., New York, New York, 1956, Chap. 13.

17. C. D. Ritchie and W. F. Sager, Prog. Phys. Org. Chem., 2, 323 (1964).

18. A. Streitwieser, Solvolytic Displacement Reactions, McGraw-Hill Book Co., Inc., New York, New York, 1962.

19. R. W. Taft and C. A. Grob, J. Amer. Chem. Soc., 96, 1236 (1974).

20. C. D. Ritchie and R. E. Uschold, J. Amer. Chem. Soc., 90, 2821 (1968).

21. D. N. Kevill, K. C. Kolwyck, D. M. Shold, and C. B. Kim, J. Amer. Chem. Soc., 95, 6022 (1973).

22. E. L. Eliel, N. L. Allinger, S. J. Angyal, and G. A. Morrison, Conformational Analysis, John Wiley and Sons, Inc., New York, New York, 1965.

23. W. J. LeNoble, Highlights of Organic Chemistry, Marcel Dekker, Inc., New York, New York, 1974, pp. 700-707. A more complete discussion is presented.

24. P. D. Bartlett and T. T. Tidwell, J. Amer. Chem. Soc., 90, 4421 (1968).

25. V. J. Shiner, R. D. Fisher, and W. Dowd, J. Amer. Chem. Soc., 91, 7748 (1969).

26. P. v. R. Schleyer, J. L. Fry, L. K. Lam, and C. J. Lancelot, J. Amer. Chem. Soc., 92, 2542 (1970).

27. V. J. Shiner, M. W. Rapp, and H. R. Pinnick, J. Amer. Chem. Soc., 92, 232 (1970).

Supplementary Reading

For a review of structural effects on solvolysis reactions, see: J. M. Harris, Prog. Phys. Org. Chem., 11, 89 (1974).

Chapter 5

STRUCTURE AND REACTIVITY: MOLECULAR ORBITAL THEORY

5-1. INTRODUCTION

The empirical relationships between structure and reactivity which we discussed in Chapter 4 are extremely useful in many cases where basic theory
offers no help. There are also many examples, however, in which the empirical relationships offer little help, but in which basic theory offers a
rational means of understanding structural effects on reactivity. For example, the fact that cyclic conjugated molecules containing 4n + 2 π-
electrons are unusually stable, and those containing 4n π-electrons are
unusually unstable, relative to their open chain analogues is easily understood on the basis of molecular orbital theory. Similarly, we shall see that
the stereospecificity of many ring forming or opening reactions is easily
rationalized in terms of molecular orbitals.

Our discussion of molecular orbital (MO) theory will begin with a simple
"how to do it" presentation of Hückel π-MO theory. The Hückel theory is
an extreme simplification of basic theory, and it is important to understand
the assumptions involved in its formulation in order to understand its limitations. It is hoped that the examples of the utility of the Hückel theory will
motivate the reader to the effort required to understand the more rigorous
discussion of molecular quantum mechanics presented in the later sections
of the chapter.

We shall assume only that the reader is acquainted with the idea that a
wavefunction Ψ specifies the state of a system in the sense that for a given
wavefunction there is associated a given energy, and that $\Psi^*\Psi$ is to be interpreted as the probability that a system (i.e., each nucleus and each
electron) is at a given point in the many dimensional space (i.e., 3 position
coordinates for each particle) of the system [1]. For an electronic wavefunction, $\Psi^*\Psi$ is proportional to the electron density at the point for which
Ψ is evaluated, and

$$\int_{-\infty}^{\infty} \Psi^*\Psi \ d\tau = 1 \qquad (5-1)$$

where $d\tau$ includes all of the coordinates of all of the electrons.

5-2. HÜCKEL π-MO RECIPE [2]

We shall see in the later development that Hückel π-MO theory is extremely easy to apply to any π-electron system, and can be stated in terms of a recipe which can be followed by anyone with access to a computer.

The recipe, stated step-by-step and illustrated with the examples of ethylene, the allyl system (CH_2CHCH_2), and the cyclopropenyl system (cyclo-C_3H_3), is as follows:

1. Draw the σ-electron structure of just the carbon skeleton of the molecule of interest, and number the carbon atoms (see example below).

Ethylene	Allyl	Cyclopropenyl
C — C 1 2	C — C — C 1 2 3	1 2 C — C \C/ 3

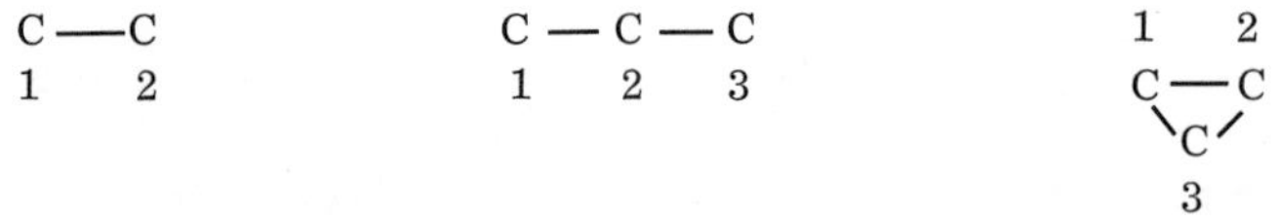

2. Write an n x n matrix H for the n atom system in which (a) all diagonal elements are 0, (b) the ij-th element of H is 1 if atom number i is directly connected to atom number j by the skeleton drawing, and (c) all other elements of H are 0:

$$H_{ethylene} = \begin{bmatrix} 0 & 1 \\ 1 & 0 \end{bmatrix}$$

$$H_{allyl} = \begin{bmatrix} 0 & 1 & 0 \\ 1 & 0 & 1 \\ 0 & 1 & 0 \end{bmatrix}$$

$$H_{cyclopropenyl} = \begin{bmatrix} 0 & 1 & 1 \\ 1 & 0 & 1 \\ 1 & 1 & 0 \end{bmatrix}$$

3. Diagonalize H with a unitary transformation. That is, find the matrix U such that

$$U^*HU = \Lambda \tag{5-2}$$

where U is a unitary matrix and Λ is a diagonal matrix.

This step is generally carried out on a computer. The H matrix for even a very large system can be diagonalized in no more than a few seconds of computer time.

The U matrices for the example systems are

$$U_{ethylene} = \begin{bmatrix} 1/\sqrt{2} & -1/\sqrt{2} \\ 1/\sqrt{2} & 1/\sqrt{2} \end{bmatrix} \quad U_{allyl} = \begin{bmatrix} 1/2 & 1/\sqrt{2} & 1/2 \\ 1/\sqrt{2} & 0 & -1/\sqrt{2} \\ 1/2 & -1/\sqrt{2} & 1/2 \end{bmatrix}$$

$$U_{cyclopropenyl} = \begin{bmatrix} 1/\sqrt{3} & 1/\sqrt{6} & 1/\sqrt{2} \\ 1/\sqrt{3} & 1/\sqrt{6} & -1/\sqrt{2} \\ 1/\sqrt{3} & -2/\sqrt{6} & 0 \end{bmatrix}$$

You should verify that U*HU = Λ for each case, and that

$$\Lambda_{ethylene} = \begin{bmatrix} 1 & 0 \\ 0 & -1 \end{bmatrix} \quad\quad allyl = \begin{bmatrix} \sqrt{2} & 0 & 0 \\ 0 & 0 & 0 \\ 0 & 0 & -\sqrt{2} \end{bmatrix}$$

$$cyclopropenyl = \begin{bmatrix} 2 & 0 & 0 \\ 0 & -1 & 0 \\ 0 & 0 & -1 \end{bmatrix} \quad\quad (5\text{-}3)$$

4. The molecular orbital energies are $\epsilon_i = \alpha + \lambda_i \beta$, where α and β are negative constants. Thus, the lowest orbital energy corresponds to the highest value of λ. The λ_i's are the elements of the diagonal matrix and are ordered in a manner corresponding to the order in which the columns of U are written. The columns of U are the molecular orbitals, in the sense that

$$\phi_i = u_{1i}\gamma_1 + u_{2i}\gamma_2 + \cdots + u_{ni}\gamma_n \quad\quad (5\text{-}4)$$

$$\phi_i = \sum_{j=1}^{n} \gamma_j u_{ji} \quad\quad (5\text{-}5)$$

where the γ's are $2p_z$ type atomic orbitals centered on the atom designated by the subscript.

The MO energies, the MO's, and a sketch of the orbitals for each of our examples are as follows:

Ethylene:

$$\epsilon_1 = \alpha + \beta \quad\quad \phi_1 = 1/\sqrt{2}\,(\gamma_1 + \gamma_2)$$

$$\epsilon_2 = \alpha - \beta \quad\quad \phi_2 = 1/\sqrt{2}\,(\gamma_1 - \gamma_2)$$

Allyl:

$$\epsilon_1 = \alpha + \sqrt{2}\,\beta \qquad\qquad \phi_1 = 1/2\,(\gamma_1 + \sqrt{2}\,\gamma_2 + \gamma_3)$$

$$\epsilon_2 = \alpha \qquad\qquad \phi_2 = 1/\sqrt{2}\,(\gamma_1 - \gamma_3)$$

$$\epsilon_3 = \alpha - \sqrt{2}\,\beta \qquad\qquad \phi_3 = 1/2\,(\gamma_1 - \sqrt{2}\,\gamma_2 + \gamma_3)$$

Cyclopropenyl:

$$\epsilon_1 = \alpha + 2\beta \qquad\qquad \phi_1 = 1/\sqrt{3}\,(\gamma_1 + \gamma_2 + \gamma_3)$$

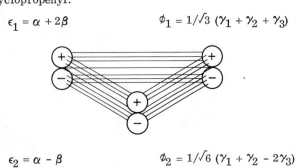

$$\epsilon_2 = \alpha - \beta \qquad\qquad \phi_2 = 1/\sqrt{6}\,(\gamma_1 + \gamma_2 - 2\gamma_3)$$

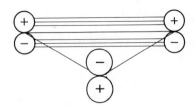

$$\epsilon_3 = \alpha - \beta \qquad\qquad \phi_3 = 1/\sqrt{2}\ (\gamma_1 - \gamma_2)$$

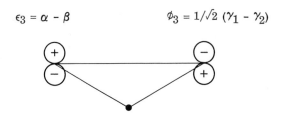

5. Each carbon atom of the π-system that we have considered has a unit positive charge. We now place electrons in the MO's according to Hund's rules to get our final system.

For the ethylene system, to get the neutral ethylene molecule we put two electrons into ϕ_1. Each electron is assigned the energy of the orbital into which it is placed. Therefore, the total π-electron energy of ethylene is $2\epsilon_1 = 2\alpha + 2\beta$.

For the allyl system, two electrons added will give us the allyl cation with π-electron energy of $2\alpha + 2\sqrt{2}\ \beta$. Addition of one more electron gives the neutral allyl radical with π-energy of $2\epsilon_1 + \epsilon_2 = 3\alpha + 2\sqrt{2}\ \beta$. Addition of another electron gives the allyl anion with π-energy of $4\alpha + 2\sqrt{2}\ \beta$.

For the cyclopropenyl system, addition of two electrons into ϕ_1 gives the cyclopropenyl cation with π-energy of $2\alpha + 4\beta$. Another electron, going into ϕ_2 gives the cyclopropenyl radical with π-energy of $3\alpha + 3\beta$. To get the cyclopropenyl anion, the next electron added must go into ϕ_3 with the same spin as the one added to ϕ_2. That is, we predict that the cyclopropenyl anion will be a triplet state, with π-energy of $4\alpha + 2\beta$.

The constant α in the above expressions is simply the energy of an electron in an isolated carbon $2p_z$ orbital. The constant β, as we see for ethylene, corresponds to the extra energy of stabilization that an electron has in a classical double bond. With these definitions, we can now talk about the resonance, or delocalization stabilization of various systems treated by the Hückel theory.

We can now say that if the allyl cation had a classical structure with a localized double bond and an empty p_z orbital, it would have a π-electron energy the same as that of ethylene, $2\alpha + 2\beta$. The Hückel theory gives the π-electron energy of $2\alpha + 2\sqrt{2}\ \beta$, and we can say that the allyl cation has a resonance stabilization of $(2\sqrt{2} - 2)\beta$ energy units.

Applying the same arguments to the allyl radical, we expect the classical structure with an isolated double bond and the odd electron in a localized p_z orbital to have an energy of $3\alpha + 2\beta$. The allyl radical, then, has a stabilization energy of $(2\sqrt{2} - 2)\beta$. Similarly, the classical structure of the allyl anion would have an isolated double bond and two electrons in a localized p_z orbital, so the allyl anion has a stabilization energy of $(2\sqrt{2} - 2)\beta$.

This reasoning gives a stabilization energy of 2β for the cyclopropenyl cation, β for the cyclopropenyl radical, and 0 for the cyclopropenyl anion.

As further examples of the use of the above recipe, the pertinent matrices and energies for the cyclic C_4 through C_7 π-systems, butadiene, and trimethylene methane are given below:

$$
\begin{array}{c}
{}^{1}C - C^{2} \\
{}_{4}C - C_{3}
\end{array}
\qquad
H =
\begin{bmatrix}
0 & 1 & 0 & 1 \\
1 & 0 & 1 & 0 \\
0 & 1 & 0 & 1 \\
1 & 0 & 1 & 0
\end{bmatrix}
\qquad
U =
\begin{bmatrix}
1/2 & 1/2 & 1/2 & 1/2 \\
1/2 & 1/2 & -1/2 & -1/2 \\
1/2 & -1/2 & -1/2 & 1/2 \\
1/2 & -1/2 & 1/2 & -1/2
\end{bmatrix}
$$

$$
=
\begin{bmatrix}
2 & 0 & 0 & 0 \\
0 & 0 & 0 & 0 \\
0 & 0 & 0 & 0 \\
0 & 0 & 0 & -2
\end{bmatrix}
$$

$\epsilon_1 = \alpha + 2\beta$ $\qquad$ $\epsilon_2 = \epsilon_3 = \alpha$ $\qquad$ $\epsilon_4 = \alpha - 2\beta$

$E^{+\cdot} = 3\alpha + 4\beta$ $\qquad$ $E^{\circ} = 4\alpha + 4\beta$ (triplet) $\qquad$ $E^{-\cdot} = 5\alpha + 4\beta$

$$
\begin{array}{c}
{}^{1}C - C^{2} \\
{}_{5}C \qquad C^{3} \\
C_{4}
\end{array}
\qquad
H =
\begin{bmatrix}
0 & 1 & 0 & 0 & 1 \\
1 & 0 & 1 & 0 & 0 \\
0 & 1 & 0 & 1 & 0 \\
0 & 0 & 1 & 0 & 1 \\
1 & 0 & 0 & 1 & 0
\end{bmatrix}
$$

$$
U =
\begin{bmatrix}
1/\sqrt{5} & 0 & \sqrt{2/5} & 0 & \sqrt{2/5} \\
1/\sqrt{5} & 0.602 & 0.195 & 0.372 & -0.512 \\
1/\sqrt{5} & 0.372 & -0.512 & -0.602 & 0.195 \\
1/\sqrt{5} & -0.372 & -0.512 & 0.602 & 0.195 \\
1/\sqrt{5} & -0.602 & 0.195 & -0.372 & -0.512
\end{bmatrix}
$$

$\epsilon_1 = \alpha + 2\beta$ $\qquad$ $\epsilon_2 = \epsilon_3 = \alpha + 0.618\beta$ $\qquad$ $\epsilon_4 = \epsilon_5 = \alpha - 1.618\beta$

$E^{+} = 4\alpha + 5.236\beta$ (triplet) $\qquad$ $E^{\circ\cdot} = 5\alpha + 5.854\beta$ $\qquad$ $E^{-} = 6\alpha + 6.472\beta$

$$H = \begin{bmatrix} 0 & 1 & 0 & 0 & 0 & 1 \\ 1 & 0 & 1 & 0 & 0 & 0 \\ 0 & 1 & 0 & 1 & 0 & 0 \\ 0 & 0 & 1 & 0 & 1 & 0 \\ 0 & 0 & 0 & 1 & 0 & 1 \\ 1 & 0 & 0 & 0 & 1 & 0 \end{bmatrix}$$

$$U = \begin{bmatrix} 1/\sqrt{6} & 0 & 1/\sqrt{3} & 0 & 1/\sqrt{3} & 1/\sqrt{6} \\ 1/\sqrt{6} & 1/2 & 1/2\sqrt{3} & 1/2 & -1/2\sqrt{3} & -1/\sqrt{6} \\ 1/\sqrt{6} & 1/2 & -1/2\sqrt{3} & -1/2 & -1/2\sqrt{3} & 1/\sqrt{6} \\ 1/\sqrt{6} & 0 & -1/\sqrt{3} & 0 & 1/\sqrt{3} & -1/\sqrt{6} \\ 1/\sqrt{6} & -1/2 & -1/2\sqrt{3} & 1/2 & -1/2\sqrt{3} & 1/\sqrt{6} \\ 1/\sqrt{6} & -1/2 & 1/2\sqrt{3} & -1/2 & -1/2\sqrt{3} & -1/\sqrt{6} \end{bmatrix}$$

$\epsilon_1 = \alpha + 2\beta \qquad \epsilon_2 = \epsilon_3 = \alpha + \beta \qquad \epsilon_4 = \epsilon_5 = \alpha - \beta \qquad \epsilon_6 = \alpha - 2\beta$

$E^\circ = 6\alpha + 8\beta$

$$H = \begin{bmatrix} 0 & 1 & 0 & 0 & 0 & 0 & 1 \\ 1 & 0 & 1 & 0 & 0 & 0 & 0 \\ 0 & 1 & 0 & 1 & 0 & 0 & 0 \\ 0 & 0 & 1 & 0 & 1 & 0 & 0 \\ 0 & 0 & 0 & 1 & 0 & 1 & 0 \\ 0 & 0 & 0 & 0 & 1 & 0 & 1 \\ 1 & 0 & 0 & 0 & 0 & 1 & 0 \end{bmatrix}$$

$$U = \begin{bmatrix} 1/\sqrt{7} & 0 & 0.535 & 0 & 0.535 & 0 & -0.535 \\ 1/\sqrt{7} & 0.418 & 0.333 & 0.521 & -0.119 & -0.232 & 0.482 \\ 1/\sqrt{7} & 0.521 & -0.119 & -0.232 & -0.482 & 0.418 & -0.333 \\ 1/\sqrt{7} & 0.232 & -0.482 & -0.418 & 0.333 & -0.521 & 0.119 \\ 1/\sqrt{7} & -0.232 & -0.482 & 0.418 & 0.333 & 0.521 & 0.119 \\ 1/\sqrt{7} & -0.521 & -0.119 & 0.232 & -0.482 & -0.418 & -0.333 \\ 1/\sqrt{7} & -0.418 & 0.333 & -0.521 & -0.119 & 0.232 & 0.482 \end{bmatrix}$$

$\epsilon_1 = \alpha + 2\beta$ $\epsilon_2 = \epsilon_3 = \alpha + 1.247\beta$ $\epsilon_4 = \epsilon_5 = \alpha - 0.445\beta$

$\epsilon_6 = \epsilon_7 = \alpha - 1.802\beta$

$E^+ = 6\alpha + 8.988\beta$ $E^\circ = 7\alpha + 8.543\beta$ $E^- = 8\alpha + 8.098\beta$ (triplet)

$$
\begin{array}{cccc}
1 & 2 & 3 & 4 \\
\end{array}
$$
$$
\text{C} - \text{C} - \text{C} - \text{C} \qquad H = \begin{bmatrix} 0 & 1 & 0 & 0 \\ 1 & 0 & 1 & 0 \\ 0 & 1 & 0 & 1 \\ 0 & 0 & 1 & 0 \end{bmatrix}
$$

$$
U = \begin{bmatrix} 0.371 & 0.600 & 0.600 & 0.371 \\ 0.600 & 0.371 & -0.371 & -0.600 \\ 0.600 & -0.371 & -0.371 & 0.600 \\ 0.371 & -0.600 & 0.600 & -0.371 \end{bmatrix}
$$

$\epsilon_1 = \alpha + 1.618\beta$ $\epsilon_2 = \alpha + 0.618\beta$ $\epsilon_3 = \alpha - 0.618\beta$ $\epsilon_4 = \alpha - 1.618\beta$

$E^\circ = 4\alpha + 4.472\beta$

$$
\begin{array}{c}
\text{C}^1 \\
| \\
\text{C}^4 \\
\end{array}
$$
$$
{}^3\text{C} \diagup \quad \diagdown \text{C}^2 \qquad H = \begin{bmatrix} 0 & 0 & 0 & 1 \\ 0 & 0 & 0 & 1 \\ 0 & 0 & 0 & 1 \\ 1 & 1 & 1 & 0 \end{bmatrix}
$$

$$
U = \begin{bmatrix} 1/\sqrt{6} & 0 & -2/\sqrt{6} & -1/\sqrt{6} \\ 1/\sqrt{6} & 1/\sqrt{2} & 1/\sqrt{6} & -1/\sqrt{6} \\ 1/\sqrt{6} & -1/\sqrt{2} & 1/\sqrt{6} & -1/\sqrt{6} \\ 1/\sqrt{2} & 0 & 0 & 1/\sqrt{2} \end{bmatrix}
$$

$\epsilon_1 = \alpha + \sqrt{3}\,\beta$ $\epsilon_2 = \epsilon_3 = \alpha$ $\epsilon_4 = \alpha - \sqrt{3}\,\beta$ $E^\circ = 4\alpha + 2\sqrt{3}\,\beta$ (triplet)

Notice that for all of the cyclic systems the MO energies occur in pairs except for the lowest and highest ones. That is, the orbitals, except for the highest and lowest, occur in degenerate pairs. A diagram of the MO energy levels for cyclic systems is shown in Figure 1.

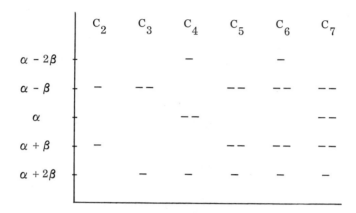

Figure 5-1. MO Energies of Cyclic Systems

It can be shown by application of group theory that this double degeneracy
of all intermediate levels for cyclic systems is general. Thus, for any
cyclic system, we may place two electrons in the lowest orbital, but then
electrons must be added four at a time to obtain a filled level (i.e., a
closed shell). This is the origin of the Hückel $4n + 2$ π-electron rule, which
helps us to understand the unusual stability of the cyclopropenium cation
($n = 0$), benzene and tropylium cation ($n = 1$), and cyclooctatetraene dianion
($n = 2$). The Hückel treatment also gives some indication of the unusual
nature of the $4n$-π-electron systems [3] in that it predicts that these systems will be ground state triplets. The treatment, however, badly over-
estimates the stabilities of such systems.

5-3. ORBITAL SYMMETRY [4]

Not only the orbital energies calculated from the Hückel theory, but also
the orbitals themselves, can help in understanding many reactions of π-
systems. In particular, the "shape," or more properly, the phase, of the
molecular orbitals determine the stereochemistry of many reactions.
Bonding between two orbitals can occur only when the orbitals overlap in
phase; that is, only when the positive lobe of one orbital overlaps in space
with the positive lobe of the other orbital, or when the negative lobe of one
overlaps with the negative lobe of the other, but not when the negative lobe
of one overlaps with the positive lobe of the other. For example, in ϕ_2 of
butadiene, since γ_2 and γ_3 have coefficients of different sign, this orbital
makes no contribution to the bonding between atoms 2 and 3. We say that
there is a node in the orbital between atoms 2 and 3. In other words, the
electron density for this orbital is zero at some point (halfway in this case)
between the two atoms.

In many cases it is not even necessary to solve the Hückel problem in order to know the phase properties of the various orbitals. The lowest orbital of a π-system is always totally bonding; that is, it has no nodes. The next lowest orbital will always have one node. For a straight chain system, this node will always be at the center of the chain. For a cyclic system, the node will bisect the ring. The nodal properties of the orbitals are associated with the energy levels of the orbitals, and we see that in all of the above examples, degenerate orbitals have the same number of nodes. For cyclic systems, then, there will always be two orbitals with one node; in the systems discussed, these orbitals are such that the nodes are at right angles to each other. For benzene, ϕ_2 has a nodal plane that passes through carbons 1 and 4; ϕ_3 has a nodal plate that bisects the C_2-C_3 and C_5-C_6 bonds.

As the orbital energies increase, so also do the number of nodes in the orbitals. A few minutes thought should convince you that this behavior arises from the requirement that the orbitals are orthogonal to each other.

In discussions of the influence of orbital phase on the pathway or stereochemistry of a reaction, it is usually found that consideration of the highest occupied MO (HOMO) and of the lowest unoccupied MO (LUMO) is sufficient. Consider, for example, the ring opening of a cyclobutene derivative:

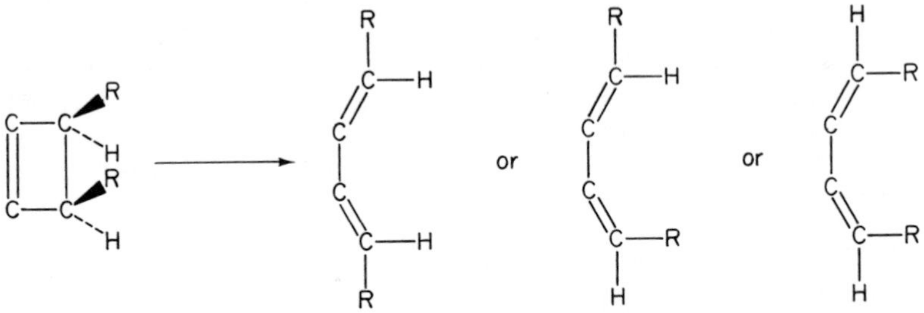

As the cyclobutene ring opens, the two electrons originally in the C-C σ-bond will begin to interact with the π-system, which is just an ethylene system. Since there are already two electrons in ϕ_1 of the ethylene bond, the only orbital available for interaction with the original σ-orbital is ϕ_2 of the ethylene system. This is the LUMO of the cyclobutene ring. There are two occupied orbitals in the π-system of the product butadiene, ϕ_1 and ϕ_2, and ϕ_2 is the HOMO of the product. This HOMO must be formed from the atomic orbitals on the carbons which formed the C-C σ-bond of the cyclobutene interacting with the LUMO of the ethylene system. A bonding overlap of the pertinent orbitals can occur only if the original sigma orbitals are rotated in the same sense:

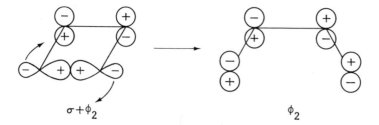

$\sigma + \phi_2$ ϕ_2

As shown in the diagram, this overlap proceeds smoothly to give ϕ_2 of the butadiene system, and forms the cis-trans product only from a cis-disubstituted reactant.

The same arguments, summarized in the diagram below, applied to the ring opening of cyclohexadiene, in which the LUMO is ϕ_3 of a butadiene system, to produce the open hexatriene, in which the HOMO must have two nodes, leads us to expect that the ring opening will proceed with opposite rotation of the two carbons of the original C-C bond:

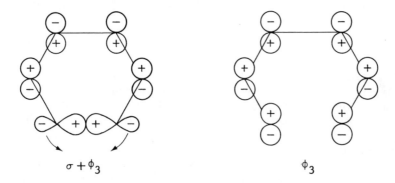

$\sigma + \phi_3$ ϕ_3

Thus, a cis-disubstituted cyclohexadiene is expected to produce either the cis-cis or the trans-trans, but not the cis-trans, product.

We discussed in earlier sections the fact that solvolyses of cyclopropyl halides or tosylates are extremely slow reactions, primarily because of the ring strain produced on going to the transition state for either an S_N1 or S_N2 reaction. Various efforts to produce the cyclopropyl cation, in fact, either fail completely or lead to the production of the allyl cation. This latter product apparently arises from a ring opening in which the pertinent orbitals are sketched below:

$$\sigma + \gamma \qquad \qquad \phi_1$$

The ring opening, then, is expected to involve opposite sense rotations of
the two carbons, leading to either a U-form or a W-form allylic cation
from a cis-disubstituted cyclopropane derivative:

U-form   W-form

The U-form, as we discussed earlier, has severe steric interactions if R
is larger than H, and we would expect the W-form to be preferred.

   In reactions of disubstituted cyclopropyl halides or tosylates, it appears
that the ring opening occurs simultaneously with the departure of the leaving group. That is, the C-C σ-bond appears to act as an intramolecular
S_N2 reagent (i.e., there appears to be neighboring group participation by
the σ-bond). For example, the acetolysis of trans,trans-2,3-dimethyl-
cyclopropyl chloride is 6500 times faster than that of the corresponding
cis,cis-isomer, and 100 times faster than that of the cis,trans-isomer.
This data is easily understandable on the basis of the symmetry of the ring
opening since backside participation of the σ-bond rotates the cis,cis-
groups toward each other [5].

   These combined effects of orbital symmetry and steric factors are even
more pronounced in the bicyclic systems

Endo   Exo

where the acetolysis of the endo-isomer is at least 10^6 faster than that of the exo-isomer since orbital symmetry and participation for the exo-isomer would lead to a very highly strained ring system with the hydrogens inside [4].

This very abbreviated discussion of some applications of Hückel MO theory should be sufficient to convince the reader of its utility. It is also important, however, to understand the limitations of the theory, and this requires a more thorough discussion of the basis of MO methods in general, and of some of the fundamental concepts of quantum mechanics.

5-4. OPERATORS AND OBSERVABLES [6]

The modern formulation of quantum mechanics is in terms of observables, operators, and state functions. A state function is some function of all of the coordinates of the system. The function is to be interpreted in terms of probabilities in the sense that F*F is a measure of the probability of finding the system with the values of the coordinates specified in the evaluation of the quantity. This interpretation places several restrictions on the type of function which can be valid. For a state function F, the restrictions are that (a) F must be continuous and single valued, and (b) the integral of F*F over all space must be finite.

We shall see in the ensuing discussion that the magnitude of F is arbitrary in the sense that αF, where α is an arbitrary constant, is a valid state function if F is a valid state function. Two conventions are frequently used in assigning a magnitude to F: either the magnitude is chosen such that F*F integrated over all space is unity, or such that it is equal to the number of particles in the system. In the following discussion of fundamentals we shall consider a one-particle system constrained to one dimension, and the two conventions are then equivalent.

One type of state function which satisfies the above restrictions is

$$F = (\frac{2\alpha}{\pi})^{1/4} \exp(-\alpha x^2) \tag{5-6}$$

It is helpful at this point to evaluate some integrals involving this function, and to introduce the Dirac notation:

$$\int_{-\infty}^{\infty} A^*(g) B \, d\tau = <A \,|\, g \,|\, B>$$

where A and B are functions of the coordinates τ, and g is some operator, such as d/dx. For the function defined in Eq. (5-6):

$$<F \,|\, F> = 1 \tag{5-7}$$

$$< F \,|\, x \,|\, F > \,=\, 0 \tag{5-8}$$

$$< F \,|\, x^2 \,|\, F > \,=\, 1/4\alpha \tag{5-9}$$

One of the most basic postulates of quantum mechanics is that for every observable of a system, there is a corresponding operator such that:

$$(\text{operator}) \ \Psi = (\text{value of observable}) \ \Psi \tag{5-10}$$

where Ψ is a state function which describes a system in a state such that only one value for the observable is possible. In general, systems will not be in such "eigenstates," and we can then only talk about average values for observables. For any given operator and corresponding observable, Eq. (5-10) has a number of solutions, Ψ_i. These exact solutions form a "complete set" of functions. This means that any arbitrary function of the coordinates of the system which meets the requirements stated above, can be expressed as a linear combination of the eigenfunctions of Eq. (5-10):

$$B = \sum_{i=1}^{\infty} \Psi_i c_i \tag{5-11}$$

If a system is in a state described by the function B, which is not an eigenstate of an operator, the "expectation value," or average value, of an observable $<m>$ with corresponding operator M, is given by

$$<B \,|\, M \,|\, B> \,=\, <m> <B \,|\, B> \tag{5-12}$$

We assume that B has been normalized, so $<B \,|\, B> = 1$.
   Substitution of Eq. (5-11) into Eq. (5-12) gives

$$\sum_{i=1}^{\infty} \sum_{j=1}^{\infty} <\Psi_i \,|\, M \,|\, \Psi_j> c_i c_j = <m> \tag{5-13}$$

Since the Ψ's are eigenfunctions of the operator M, then

$$M\Psi_i = m_i \Psi_i \tag{5-14}$$

where the m_i's are eigenvalues of the operator M.
   Substitution of Eq. (5-14) into (5-13) gives

$$\sum_{i=1}^{\infty} \sum_{j=1}^{\infty} <\Psi_i \,|\, \Psi_j> c_i c_j m_j = <m> \tag{5-15}$$

Equations (5-10) and (5-14) have the property that their solutions, Ψ_i either are automatically, or can be made to be orthogonal to one another:

$$<\Psi_i|\Psi_j> = 0 \quad \text{if } i \neq j \tag{5-16}$$

Equation (5-16), along with the requirement already stated that

$$<\Psi_i|\Psi_i> = 1 \tag{5-17}$$

substituted into Eq. (5-15) gives

$$\sum_{j=1}^{\infty} c_j^2 \, m_j = <m> \tag{5-18}$$

Now, since the function B was normalized

$$<B|B> = \sum_{i=1}^{\infty} \sum_{j=1}^{\infty} <\Psi_i|\Psi_j> c_j c_i = \sum_{j=1}^{\infty} c_j^2 = 1 \tag{5-19}$$

and it follows that $<m>$ is a weighted average of the eigenvalues m_j and can be no smaller than the smallest m_j. We shall see that this conclusion becomes extremely important when we begin to consider the energy of a system associated with an approximate wave function.

Thus far, our discussion of Eqs. (5-10) - (5-19) has been in very general terms. Let us now consider some specific examples of observables and operators.

The following observables and corresponding operators are the ones of interest for our purposes:

Observable	Operator
Coordinate: q_i	q_i
Momentum: $m_i v_i = u_i$	$-i\hbar \dfrac{\partial}{\partial q_i} \; (\hbar = \dfrac{\hbar}{2\pi})$
Kinetic energy: $\dfrac{1}{2m_i} v_i^2 = \dfrac{1}{2m_i} u_i^2$	$-\dfrac{1}{2m_i} \hbar^2 \dfrac{\partial^2}{\partial q_i^2}$

Note that the expression for the kinetic energy operator is derived by substituting the momentum operator into the classical expression for the observable. This is quite general and allows us to formulate operators for any observable for which we can write an expression in terms of coordinates and momenta of the system.

As examples of the use of these operators, let us examine some of the properties of our one-dimensional one-particle system described by the function F in Eq. (5-6). The expectation value for the position x of the system is

$$<F|x|F> = <x> = 0 \qquad\qquad (5\text{-}20)$$

This equation says that the average position of the particle is at the origin of the coordinate system; that is, it is just as often at a negative coordinate as at a positive coordinate.

We can also determine the average displacement of the particle. A measure of the average displacement would be the average value of x^2, which is given by

$$<F|x^2|F> = <x^2> = 1/4\alpha \qquad\qquad (5\text{-}21)$$

The root mean square displacement is then $(1/4\alpha)^{1/2}$.

The average momentum of the particle is

$$<F|-i\hbar\frac{\partial}{\partial x}|F> = i\hbar(2\alpha)<F|x|F> = <u> = 0 \qquad\qquad (5\text{-}22)$$

The average of the square of the momentum is

$$<F|-\frac{\hbar^2\partial^2}{\partial x^2}|F> = 2\alpha\hbar^2<F|\frac{\partial}{\partial x}|xF>$$

$$= 2\alpha\hbar^2 <F|F - 2\alpha x^2 F> = 2\alpha\hbar^2<F|F> - 4\alpha^2\hbar^2<F|x^2|F>$$

$$= 2\alpha\hbar^2 - 4\alpha^2\hbar^2\frac{1}{4\alpha} = \alpha\hbar^2 = <u^2> \qquad\qquad (5\text{-}23)$$

and the root mean square momentum is $\hbar\alpha^{1/2}$.

The kinetic energy of the particle is

$$\frac{-\hbar^2}{2m} <F|\frac{\partial^2}{\partial x^2}|F> = (1/2m)<u^2> = \frac{\alpha\hbar^2}{2m} = <T> \qquad\qquad (5\text{-}24)$$

Let us now suppose that the particle under consideration is operated on by Hooke's law force, given by

Force = kx

$$V = \int kx\,dx = \frac{1}{2}kx^2 \qquad\qquad (5\text{-}25)$$

where k is the Hooke's law constant and V is the potential energy of the particle. For our particle described by the function F then, the expectation value for the potential energy is

$$<V> = <F \,|\, \frac{1}{2} kx^2 \,|\, F> = \frac{k}{2} <F\,|\,x^2\,|\,F> = \frac{k}{2} <x^2> = \frac{k}{8\alpha} \tag{5-26}$$

The expectation value for the total energy of the particle is simply the sum of the expectation values of the kinetic and potential energies, so that

$$<E> = <T> + <V> = \frac{\alpha \hbar^2}{2m} + \frac{k}{8\alpha} \tag{5-27}$$

We may now note that F could be an eigenfunction of the total energy operator, $H = T_{op} + V_{op}$, if the value of α were properly chosen. Writing the operator equation

$$HF = (T_{op} + V_{op})F = \frac{-\hbar^2}{2m} \frac{\partial^2 F}{\partial x^2} + \frac{k}{2} x^2 F$$

$$= \frac{-\hbar^2}{2m} (-2\alpha F + 4\alpha^2 x^2 F) + \frac{kx^2}{2} F$$

$$= \frac{\alpha \hbar^2}{m} F + (\frac{k}{2} - \frac{2\alpha^2 \hbar^2}{m}) x^2 F$$

we see that if the term in $x^2 F$ were to vanish, then F would be an eigenfunction of H with eigenvalue of $\alpha \hbar^2/m$. The offending term can be made to vanish if we set

$$\frac{k}{2} - \frac{2\alpha^2 \hbar^2}{m} = 0$$

$$\therefore \alpha = \frac{(km)^{1/2}}{2\hbar}$$

The eigenvalue then becomes $(\hbar/2)\,(k/m)^{1/2}$.

Although the above example has been presented primarily to give some practical experience with expectation value, eigenfunction, and eigenvalue calculations, it is actually part of the problem of the one-dimensional harmonic oscillator encountered in discussions of molecular vibrations. We will have occasion to refer back to this example when we discuss isotope effects on reaction rates.

5-5. MOLECULAR ORBITALS [7]

More importantly for the present purposes, we are now in a position to develop the application of quantum mechanics to the problem of wavefunctions of molecules. We will begin this development by considering the electronic wavefunction for a molecule consisting of N nuclei and n electrons. The Hamiltonian operator (i.e., the total energy operator) for the electrons is

$$H = -\frac{1}{2} \sum_{i=1}^{n} \left(\frac{\partial^2}{\partial x_i^2} + \frac{\partial^2}{\partial y_i^2} + \frac{\partial^2}{\partial z_i^2} \right)$$

$$-\sum_{i=1}^{n} \sum_{j=1}^{N} \frac{Z_j}{r_{ij}} + \sum_{i=1}^{n-1} \sum_{j=i+1}^{n} \frac{1}{r_{ij}} \tag{5-28}$$

in which the first summation is over all electrons, each with cartesian coordinates x_i, y_i, and z_i; the second summation is the coulombic attractions of the nuclei and electrons, with each nucleus having a charge of Z_j and the distance between the i-th electron and the j-th nucleus being r_{ij}; and the third summation counts the coulombic repulsions of the electrons, with the dummy variables i and j limited as shown so that we count each pairwise repulsion only once. Equation (5-28) is written in units of a.u. (atomic units), in which the unit of mass is the mass of an electron divided by $\hbar^2$, the unit of charge is the charge of an electron, and the unit of length is the Bohr radius, 5.282×10^{-9} cm. This is the reason that m, for example, does not appear in the equation. The unit of energy, 1 a.u., is equal to 627 kcal/mole in this system of units.

We shall assume at the outset that the wavefunction for any molecule can be written as the product of an electronic function and a nuclear function, each of which satisfy the appropriate operator equations. This is the Born-Oppenheimer approximation, and leads to negligible errors for nearly all molecules. We can, therefore, write the Schrödinger equation for the electronic wavefunction,

$$H\Psi_i = E_i\Psi_i \tag{5-29}$$

where H is defined in Eq. (5-28).

It must be understood that an exact solution of a differential equation of the complexity of Eq. (5-29) for any molecule of interest to an organic chemist is just not possible with present theory. We must content ourselves with approximate solutions.

The basis of all approximate solutions of Eq. (5-29) lies in Eq. (5-18). Our procedure will be to guess some form for a wavefunction, in which there will be some adjustable parameters [just as α was an adjustable parameter in the one-particle function of Eq. (5-6)]. From Eq. (5-18), we know that the expectation value for the energy which we calculate from our guessed function can never be less than the lowest eigenfunction of Eq. (5-29), which is the ground state energy of the molecule. Thus, we can vary our adjustable parameters to make $<E>$ as small as possible for the form of our guessed function.

Remember, however, that we are not completely free in guessing a form for the wavefunction; it must satisfy the requirements stated on p. 139, above, and, in addition, it must satisfy the Pauli exclusion principle. We might hope at the outset that a wavefunction for our n-electron system can be written as a product of one-electron terms:

$$\Psi_{guessed} = \phi_1(1) \; \phi_2(2) \; \phi_3(3) \; \ldots \; \phi_n(n) \tag{5-30}$$

where each of the ϕ's contains only the coordinates of the electron designated by the number in parentheses following it. We can easily anticipate, however, what would happen to such a function if we made all of the ϕ's of the same form and attempted to minimize $<E>$ with respect to adjustable parameters: all of the electrons would be preferentially placed in the same region of low potential close to the nuclei, thereby violating the Pauli exclusion principle. This deficiency of Eq. (5-30) can be rectified only if we require that the wavefunction be antisymmetric with respect to the interchange of any two electrons. That is, we must require that

$$P_{ij}\Psi = -\Psi \tag{5-31}$$

where P_{ij} is an operator that permutes the coordinates of electron i and electron j. This condition is sufficient to ensure that each of the ϕ's is an independent function in the sense that

$$\phi_j \neq \sum_{i \neq j} \phi_i c_i \tag{5-32}$$

It is then fairly easy to show that the requirement:

$$<\phi_i(1) \, | \, \phi_j(1)> = 0 \quad \text{if } i \neq j \tag{5-33}$$

which we shall see is a convenience, does not place any further limitation on the form of Ψ.

A simple and convenient way to write a function satisfying Eq. (5-31) and which is expressed as a product of one-electron functions, ϕ, is as a determinant:

$$\Delta = \begin{vmatrix} \phi_1(1) & \phi_2(1) & \phi_3(1) & \cdots & \phi_n(1) \\ \phi_1(2) & \phi_2(2) & \phi_3(2) & \cdots & \phi_n(2) \\ \phi_1(3) & \phi_2(3) & \phi_3(3) & \cdots & \phi_n(3) \\ \cdot & \cdot & \cdot & \cdots & \cdot \\ \phi_1(n) & \phi_2(n) & \phi_3(n) & \cdots & \phi_n(n) \end{vmatrix} \tag{5-34}$$

This form contains all of the requirements of Eq. (5-31), since the rules of determinants state that (1) the interchange of any two rows or columns of a determinant changes the sign of the determinant, and (2) if any two rows or columns of a determinant are identical, or differ only by a constant factor, the determinant vanishes. You should be able to show now that

$$<\Delta|\Delta> = n!<\phi_i|\phi_i> = n! \tag{5-35}$$

if

$$<\phi_i|\phi_i> = 1 \quad \text{for all } i.$$

Thus, a guessed wavefunction

$$\Psi_{guessed} = (1/n)^{1/2}\Delta \tag{5-36}$$

in which the factor $(1/n)^{1/2}$ assures that $<\Psi|\Psi> = 1$ if the ϕ's are orthonormal, is the simplest type of function with which we may attempt to approximate the true wavefunction for an n-electron system. A function of this form is called a Slater determinant.

We have already introduced a serious approximation in Eq. (5-36). There is no possibility in a function of this form that two electrons could correlate their motions, since we have written the coordinates as independent functions. In the exact wavefunction, there is certainly the possibility for motions to be correlated, and this correlation will lower the energy of the system since it will reduce electron-electron repulsions. The error introduced in $<E>$ as a result of the form of Eq. (5-36) is called the correlation energy. It is the difference between the true energy of the system in its ground state and the best possible expectation value which could be calculated from a single Slater determinantal wavefunction.

The one-electron functions ϕ which we have written in the above equations must contain all of the coordinates of an electron. That is, the coordinate designation must include a "spin coordinate" for the electron, as well as the space coordinates. Thus, it is possible for ϕ_1 and ϕ_2, for example, in Eq. (5-34) to be of the form

$$\phi_1 = \phi_1' \, \lambda^+$$

$$\phi_2 = \phi_1' \, \lambda^- \qquad\qquad\qquad\qquad\qquad (5\text{-}37)$$

where the ϕ's on the right side of the equations contain only space coordinates, and the λ's contain only spin coordinates and satisfy the equations

$$\langle\lambda^+|\lambda^+\rangle = \langle\lambda^-|\lambda^-\rangle = 1$$

$$\langle\lambda^+|\lambda^-\rangle = \langle\lambda^-|\lambda^+\rangle = 0 \qquad\qquad\qquad (5\text{-}38)$$

The adoption of Eqs. (5-37) and (5-38) for a closed shell system, in fact, imposes no further approximation for the expectation value of the energy. It considerably reduces the work, since we now need only consider n/2 space orbitals.

In order to proceed with our development, we must now tackle the messy problem or expanding $\langle\Psi|H|\Psi\rangle$ in terms of the ϕ's. This is no trivial task, but the result is fairly simple.

We can begin the expansion by writing H as a sum of one-electron operators H_i and two-electron operators H_{ij}:

$$H_i = -\frac{1}{2}\left[\frac{\partial^2}{\partial x_i^2} + \frac{\partial^2}{\partial y_i^2} + \frac{\partial^2}{\partial z_i^2}\right] - \sum_{k=1}^{N} \frac{z_k}{r_{ik}}$$

$$H_{ij} = \frac{1}{r_{ij}} \qquad\qquad\qquad\qquad\qquad (5\text{-}39)$$

On substitution into Eq. (5-28), we get

$$H = \sum_{i=1}^{n} H_i + \sum_{i=1}^{n-1} \sum_{j>i}^{n} H_{ij} \qquad\qquad\qquad (5\text{-}40)$$

The equation for the expectation value of the energy of the system is

$$\langle E \rangle = \langle\Psi|H|\Psi\rangle = \sum_{i=1}^{n} \langle\Psi|H_i|\Psi\rangle + \sum_{i<j} \langle\Psi|H_{ij}|\Psi\rangle \qquad (5\text{-}41)$$

Now, since H_i operates on the coordinates of only electron i, the terms in the first summation, on expansion of Ψ will be of two types, depending upon whether electrons other than i are in identical orbitals or different orbitals in the two parts of the product function. For example, a term such as

$$<\phi_1(2)\ \phi_2(1)\ \phi_3(3)\ \cdots\ \phi_n(n)\,|H_1|\,\phi_1(2)\ \phi_2(1)\ \phi_3(3)\ \cdots\ \phi_n(n)>$$

$$= <\phi_2(1)\,|H_1|\,\phi_2(1)> <\phi_1(2)\,|\phi_1(2)> \cdots <\phi_n(n)\,|\phi_n(n)>$$

in which the term on the left of H_1 and the term on the right of H_1 are identical, has the nonvanishing value shown. If, however, any two electrons were permuted in the left term but not permuted in the right term, Eq. (5-33) requires that the total term vanish. There will be $(n-1)!$ nonvanishing terms for each of the H_i:

$$<\Delta\,|H_i|\,\Delta> = \sum_{j=1}^{n} <\phi_j(i)\,|H_i|\,\phi_j(i)> (n-1)!$$

Since there are n H_i's, and since

$$<\phi_j(i)\,|H_i|\,\phi_j(i)> = <\phi_j(k)\,|H_k|\,\phi_j(k)>$$

it follows that

$$\sum_{i=1}^{n} <\Psi\,|H_i|\,\Psi> = \sum_{i=1}^{n} <\Delta\,|H_i|\,\Delta> (1\,|n!) = \sum_{i=1}^{n} <\phi_i\,|H_1|\,\phi_i> \qquad (5\text{-}42)$$

where the ϕ's are of the form given in Eq. (5-37), we may further write

$$\sum_{i=1}^{n} <\Psi\,|H_i|\,\Psi> = 2 \sum_{i=1}^{n/2} <\phi'_i(1)\,|H_1|\,\phi'_i(1)> \qquad (5\text{-}43)$$

where the last summation is over the space orbitals only.

The expansion of the two-electron operator term in Eq. (5-41) is somewhat more difficult, but given enough time and paper, you should be able to apply the same type of reasoning as above to arrive at the result

$$\sum_{i<j} <\Psi\,|H_{ij}|\,\Psi> = \sum_{i=1}^{n/2} \sum_{j=1}^{n/2} \left[2<\phi'_i(1)\ \phi'_j(2)\,|\frac{1}{r_{12}}\,|\phi'_i(1)\ \phi'_j(2)> \right.$$
$$\left. - <\phi'_i(1)\ \phi'_i(2)\,|\frac{1}{r_{12}}\,|\phi'_j(1)\ \phi'_j(2)> \right] \qquad (5\text{-}44)$$

where the summations on the right hand side are over space orbitals ϕ'.

5-6. LINEAR COMBINATIONS OF ATOMIC ORBITALS (LCAO-MO) [7]

In nearly all treatments of molecules, the molecular orbitals are expressed as linear combinations of atomic orbitals (LCAO). The atomic orbitals are said to form a "basis set" for the molecular orbitals. Two types of basis functions are commonly used, the Slater type orbitals (STO);

$$\gamma_i = Y_i(\theta,\phi) \exp(-\alpha_i r)$$

or the Gaussian type orbitals (GTO);

$$\gamma_i = Y_i(\theta,\phi) \exp(-\alpha_i r^2)$$

where r, θ, and ϕ are spherical coordinates centered on a particular atomic nucleus. The STO's are the familiar hydrogenlike 1s, 2s, 2p, etc. atomic orbitals. The GTO's are similar types of functions which are easier to work with than STO's because of ease of integral evaluation, but which do not have the sharp peak at the nucleus characteristic of the exponential part of the STO's.

The basis functions for a calculation are simply a mathematical device, and it should come as no surprise that several of each type of atomic orbital on each nucleus are required for accurate calculations, and that 3d type functions are used on first-row atoms in the best calculations. The situation is quite analogous to a Fourier analysis in that any function can be expressed in terms of other functions if enough of them are used. If one used an infinite number of basis functions, it would be possible to express any determinantal wavefunction in terms of them. In actual practice, of course, one must settle for a finite basis set and be willing to accept some error due to this limitation. For example, a calculation on the methane molecule using 50 basis functions, carefully chosen, falls short of the best possible single determinant wavefunction by several kcals as measured by the expectation value of the calculated energy.

For the present, let us assume that we have chosen some number, say p, of basis functions γ, with which we shall express the molecular orbitals:

$$\phi'_j = \sum_{i=1}^{p} \gamma_i C_{ij} \tag{5-45}$$

The coefficients C_{ij} are to be chosen in such a way that the expectation value of the energy of the system is minimized. From Eqs. (5-18) and (5-19) we know that the best value calculated must be greater than or equal to the true ground state energy of the system.

The first step in the minimization procedure is to expand Eq. (5-41) in terms of the basis functions according to Eq. (5-45):

$$<E> = \sum_{i=1}^{n/2} \sum_{k=1}^{p} \sum_{l=1}^{p} 2C_{ki}C_{li} <\gamma_k |H_1| \gamma_l>$$

$$+ \sum_{i=1}^{n/2} \sum_{j=1}^{n/2} \sum_{k=1}^{p} \sum_{l=1}^{p} \sum_{m=1}^{p} \sum_{q=1}^{p} C_{ki}C_{lj}C_{mi}C_{qj} [2<km|lq>$$

$$- <kl|mq>] \qquad\qquad (5\text{-}46)$$

where we have used the shorthand notation

$$< km | lq > = < \gamma_k(1) \, \gamma_l(2) | 1/r_{12} | \gamma_m(1) \, \gamma_q(2)>$$

and where the indices i and j run over the occupied orbitals, and the indices k, l, m, and q run over the basis functions.

We may simplify the appearance of Eq. (5-46) by noting that the indices i and j do not appear in the integrals, but only in the coefficients. We may therefore define

$$D_{kl} = \sum_{i=1}^{n/2} C_{ki}C_{li} \qquad\qquad (5\text{-}47)$$

where, again, the index i runs over occupied orbitals. Note that if we define a matrix, C', having n/2 columns and p rows as follows:

$$C' = \begin{bmatrix} C_{11} & C_{12} & \cdot & \cdot & \cdot & C_{1n/2} \\ C_{21} & C_{22} & \cdot & \cdot & \cdot & C_{2n/2} \\ \cdot & & \cdot & \cdot & \cdot & \cdot \\ C_{p1} & C_{p2} & \cdot & \cdot & \cdot & C_{pn/2} \end{bmatrix}$$

the matrix D having elements D_{kl}, is defined by the following matrix equation:

$$D = C'C'^* \qquad\qquad (5\text{-}48)$$

The matrix D is called the density matrix for reasons that we shall discuss later.

With these definitions, Eq. (5-46) becomes

$$<E> = \sum_{k=1}^{p} \sum_{l=1}^{p} 2D_{kl} <\gamma_k | H_1 | \gamma_l>$$

$$+ \sum_{k=1}^{p} \sum_{l=1}^{p} \sum_{m=1}^{p} \sum_{q=1}^{p} D_{km}D_{lq} [2<km|lq> - <kl|mq>] \qquad (5\text{-}49)$$

In order to minimize the expectation value of the energy, we must set the partial derivative with respect to each coefficient C_{rs} equal to 0, subject to the condition that the orbitals remain orthonormal. This latter condition is introduced by means of LaGrange's undetermined multipliers, ϵ_i [10]

$$\frac{\partial <E>}{\partial C_{rs}} - 2 \sum_{i=1}^{n/2} \frac{\epsilon_i \partial <\phi_i|\phi_i>}{\partial C_{rs}} = 0 \qquad (5\text{-}50)$$

for each r and for each s.

From Eq. (5-47) we find

$$\frac{\partial D_{rl}}{\partial C_{rs}} = C_{ls} \qquad (5\text{-}51)$$

Carrying out the operations indicated in Eq. (5-50), using Eq. (5-49) for $<E>$ and with a lot of algebra, we obtain

$$\frac{\partial <E>}{\partial C_{rs}} = \sum_{k=1}^{p} 4C_{ks} <\gamma_r | H_1 | \gamma_k>$$

$$+ \sum_{k=1}^{p} \sum_{\substack{l=1 \\ m \geq 1}}^{p} 4D_{lm}C_{ks} [4<rk|lm> - <rl|km> - <rm|kl>] \qquad (5\text{-}52)$$

where we have made use of the identities $<ik|jl> = <ki|jl> = <ik|lj> = <jl|ki>$, etc. in collecting terms.

The second part of the differentiation indicated in Eq. (5-50) proceeds similarly to give

$$2 \sum_{i=1}^{n/2} \epsilon_i \partial \frac{<\phi_i|\phi_i>}{\partial C_{rs}} = \epsilon_s \sum_{k=1}^{p} 4C_{ks} <\gamma_r | \gamma_k> \qquad (5\text{-}53)$$

To further simplify the writing of equations, let us define

$$I_{kl} = <\gamma_k|H_1|\gamma_l> \tag{5-54a}$$

$$S_{kl} = <\gamma_k|\gamma_l> \tag{5-54b}$$

$$R_{kl} = \sum_{\substack{m=1 \\ q \geq m}}^{p} D_{mq} [4<kl|mq> - <km|lq> - <kq|lm>] \tag{5-54c}$$

$$F_{kl} = I_{kl} + R_{kl} \tag{5-54d}$$

Then, substitution of Eqs. (5-52) and (5-53) into Eq. (5-50) gives

$$\sum_{k=1}^{p} I_{rk}C_{ks} + \sum_{k=1}^{p} R_{rk}C_{ks} = \epsilon_s \sum_{k=1}^{p} S_{rk}C_{ks}$$

or

$$\sum_{k=1}^{p} F_{rk}C_{ks} = \epsilon_s \sum_{k=1}^{p} S_{rk}C_{ks} \tag{5-55}$$

for all r and all s. The entire set of equations may be more simply written in matrix notation:

$$FC = SC\epsilon \tag{5-56}$$

subject to the condition that $C*SC = 1$, and where the elements of F, C, and S are defined in Eqs. (5-47) and (5-54), and the matrix ϵ is a diagonal matrix. Equation (5-56) is called the Roothan equation and is used in nearly all calculations involving molecules. The use of Eq. (5-56), with the development presented, is called the Self-Consistent Field (SCF) LCAO-MO method.

Given F and S, which are evaluated by actually carrying out the integrations indicated in the defining equations, the solution of Eq. (5-56) is a matrix diagonalization problem. This is very simply seen if the basis set if chosen such that the S matrix is the identity matrix (i.e., if the basis set is an orthonormal set of functions). In this case the matrix C must be a unitary matrix if the molecular orbitals are to be orthonormal, and, since F is a Hermitean matrix, the problem is the common one of diagonalizing a matrix by a unitary transformation:

$$C*FC = \epsilon \tag{5-57}$$

If the basis functions are not orthonormal, and usually they are not, an extra step must be inserted in which the S matrix is transformed to the identity matrix. This can always be done, since S is Hermitean and has all positive–definite eigenvalues. The steps in the transformation are as follows:

1. Find the unitary matrix U which diagonalizes S:

$$U^*SU = \Lambda \tag{5-58}$$

2. Define the matrix $\Lambda^{-1/2}$ such that:

$$\Lambda^{-1/2}\Lambda\,\Lambda^{-1/2} = I \tag{5-59}$$

The elements of $\Lambda^{-1/2}$ are simply defined from the elements of Λ:

$$\lambda_i^{-1/2} = 1/\sqrt{\lambda}_i$$

3. Define the matrix G:

$$G = U\Lambda^{-1/2}$$

from which it follows that

$$G^{-1} = \Lambda^{1/2}U^* \qquad G^* = \Lambda^{-1/2}U^* \qquad (G^*)^{-1} = (G^{-1})^* \qquad G^*SG = I$$

Then operating on Eq. (5-56),

$$G^*FC = G^*SGG^{-1}C\,\epsilon \qquad G^*FG(G^{-1})C = (G^{-1}C)\,\epsilon$$

4. Define $G^{-1}C$:

$$G^{-1}C = M \quad \text{(note that M is unitary, and that } C = GM)$$
$$G^*FG = F'$$

From this we get an equation of the same form as (5-57):

$$M^*F'M = \epsilon \tag{5-60}$$

Equation (5-60) is again just the matrix diagonalization problem. In practice, one evaluates the matrix F', finds the unitary matrix M which diagonalizes it, and then transforms M to C as indicated in the above equations.

The self-consistency feature of Eq. (5-56) comes as a result of the fact that one must know the density matrix for the system [i.e., Eq. (5-48)] before the F matrix can be evaluated. One proceeds by first guessing a density matrix (for example, we could set all elements equal to 0), evaluating F, and then obtaining C. This C matrix would then be used to re-evaluate F and the procedure repeated until the C matrix used to set up F is the same one which solves Eq. (5-56).

The real problem in carrying out molecular calculations arises from the fact that on the order of p^4 integrals must be evaluated if p basis functions are used. This severely limits the size of molecules which can be handled in a straightforward application of the SCF-LCAO-MO method. Even with a very fast, high capacity computer, molecules much larger than ethane require unrealistically long times for reasonably accurate calculations.

It is primarily for this reason that the various "semiempirical" MO schemes have been developed [8]. All such schemes avoid the large number of integral evaluations by approximating the values of some or all of the integrals in Eq. (5-54). Usually some of the integrals are neglected completely, and others are evaluated from empirical data. In an important sense, many of the schemes simply become elaborate parameterization methods for fitting experimental data and retain little of the basic quantum mechanical derivation.

5-7. HÜCKEL MO THEORY

The Hückel theory represents the ultimate approximation in evaluating integrals when applied to π-electron systems. In this scheme, one starts with the assumption that the σ-electron system of a molecule provides only an "electron atmosphere" in which the π-electrons move, and that this atmosphere is completely independent of the π-orbitals. This allows the application of Eq. (5-56) to the π-orbitals only. A specific basis set for the calculations is not formally written, but is visualized to consist of one $2p_z$ atomic orbital centered on each atom of the π-system. The further approximations eliminate the need to write the form of these functions.

One of the further approximations is that the basis functions form an orthonormal set, which allows application of Eq. (5-57). The final approximations concern the evaluation of the F matrix and eliminate the need to iterate to self-consistency. The elements of F are simply approximated as follows:

$F_{ii} = \alpha$ for all i.   (5-61)

$F_{ij} = \beta$ if orbital i is centered on an atom bonded by the σ-bonds to the
          atom on which orbital j is centered.   (5-62)

$F_{ij} = 0$ otherwise.

The diagonal elements F_{ii} are visualized to have a value α corresponding to the energy of an electron in a $2p_z$ orbital on a carbon in the "electron atmosphere" of the σ-system, but otherwise isolated from other nuclei. The nonzero off-diagonal elements F_{ij} are visualized to have a value β corresponding to the "delocalization energy" of an electron in a localized two-orbital π-bond, such as in ethylene. Both α and β are negative numbers since the reference energy is always that of electrons and nuclei separated to infinite distances.

From Eqs. (5-61) and (5-62), the matrix F always has α's at all positions along the principal diagonal, and has either β's or 0's in off-diagonal positions. We can now show that the solution of Eq. (5-57) involves the steps which were given earlier on pp. 128-131 of this chapter.

Suppose we have the Hückel matrix, F, set up according to Eqs. (5-61) and (5-62), and we multiply this matrix by the constant $1/\beta$ and then subtract from the resulting matrix α/β times the identity matrix:

$$(1/\beta)F - (\alpha/\beta)I = H \qquad (5\text{-}63)$$

The resulting matrix H contains 1's in all positions where the matrix F contained β's, and 0's in all other positions. Thus, H is just the matrix which is described in step 2 on p. 128. We now show, step-wise, that any unitary matrix C which diagonalizes H also diagonalizes F.

Assume that C has been found such that

$$C^*HC = \Lambda \qquad (5\text{-}64)$$

Left-multiplication of Eq. (5-63) by C^* and right-multiplication by C gives

$$C^*HC = C^*[(1/\beta)F - (\alpha/\beta)I]C$$

and by the distributive law for matrix multiplication,

$$C^*HC = C^*(1/\beta)FC - C^*(\alpha/\beta)IC$$

Then, since any constant commutes with any matrix,

$$C^*HC = (1/\beta)C^*FC - (\alpha/\beta)C^*IC$$

and, since C is a unitary matrix,

$$C^*HC = (1/\beta)C^*FC - (\alpha/\beta)I = \Lambda$$

Now, since Λ and $(\alpha/\beta)I$ are both diagonal matrices, it follows that $(1/\beta)$ C^*FC must also be diagonal:

$$C^*FC = \beta \Lambda + \alpha I = \epsilon \tag{5-65}$$

Also, the diagonal matrix $C^*FC = \epsilon$ has elements which are related to the elements of Λ by the equation

$$\epsilon_i = \alpha + \beta \lambda_i \tag{5-66}$$

This is the equation used in step 4 on p. 129.

The construction of the molecular orbitals according to the discussion on p. 128 and 130 follows from the derivation of Eq. (5-57). The columns of C are arranged in an order corresponding to the order of the orbital energies in ϵ.

5-8.   THE FIRST-ORDER DENSITY MATRIX [9]

We shall close our discussion of MO theory with a consideration of electron densities and bond orders of molecules which can be obtained from the molecular orbitals.

Suppose that we have solved Eq. (5-56) for some molecule and obtained the occupied orbitals ϕ'_1, ϕ'_2, $\ldots$, $\phi'_{n/2}$, with two electrons in each orbital. The total electron density of the system is then

$$\text{Total density} = \sum_{i=1}^{n/2} 2 <\phi'_i | \phi'_i> \tag{5-67}$$

which, on expansion with Eq. (5-45) gives

$$\text{Total density} = 2 \sum_{i=1}^{n/2} \sum_{j=1}^{p} \sum_{k=1}^{p} c_{ji} c_{ki} <\gamma_j | \gamma_k> \tag{5-68}$$

Using the density matrix defined in Eq. (5-48),

$$\text{Total density} = 2 \sum_{j=1}^{p} \sum_{k=1}^{p} D_{jk} <\gamma_j | \gamma_k> \tag{5-69}$$

This equation may be interpreted as stating that the total electron density of the molecule is the sum of contributions due to (a) "electron occupancy" of each of the basis functions, γ_j, contributing $2D_{jj} <\gamma_j | \gamma_j>$, and (b) "electron sharing" between pairs of basis functions, γ_j and γ_k, each contributing $2D_{jk} [<\gamma_j | \gamma_k> + <\gamma_k | \gamma_j>]$. Since these basis functions are centered on individual nuclei, we can visualize the contributions as either being centered on particular nuclei or being shared between pairs of nuclei.

The total of the contributions centered on any particular nucleus r is called the "center population," q_r, and is defined as follows:

$$q_r = \sum_k \left\{ \sum_l 2D_{kl} <\gamma_k | \gamma_l> + \sum_m 2D_{km} <\gamma_k | \gamma_m> \right\} \qquad (5\text{-}70)$$

where the indices k and l are over the orbitals centered on nucleus r, and the index m is over orbitals centered on all nuclei other than r. The total of the contributions shared between any two nuclei, r and s, is called the "bond order," b_{rs}, defined as follows:

$$b_{rs} = \sum_i \sum_j 4D_{ij} <\gamma_i | \gamma_j> \qquad (5\text{-}71)$$

where the index i is over all orbitals centered on nucleus r, and the undex j is over those centered on nucleus s.

In the Hückel approximation, the center populations become simply $q_r = 2D_{rr}$ and the bond orders are formally 0, since we have assumed that the basis set is orthonormal. We can visualize relaxing the orthogonality condition, however, and the bond orders would then be simply related to the off-diagonal elements of the density matrix. These off-diagonal elements are sometimes referred to as the "mobile π-electron bond orders."

As specific illustrations of these concepts as applied to Hückel theory, we can calculate the density matrices for ethylene, allyl anion, and benzene from the C's given on p. 129 and 133.

$$D_{\text{ethylene}} = \begin{bmatrix} 1/\sqrt{2} \\ 1/\sqrt{2} \end{bmatrix} [1/\sqrt{2} \quad 1/\sqrt{2}] = \begin{bmatrix} 1/2 & 1/2 \\ 1/2 & 1/2 \end{bmatrix}$$

Thus, $q_1 = q_2 = 1$ and bond order $\propto 1/2$.

$$D_{\text{allyl anion}} = \begin{bmatrix} 1/2 & 1/\sqrt{2} \\ 1/\sqrt{2} & 0 \\ 1/2 & -1/\sqrt{2} \end{bmatrix} \begin{bmatrix} 1/2 & 1/\sqrt{2} & 1/2 \\ 1/\sqrt{2} & 0 & -1/\sqrt{2} \end{bmatrix}$$

$$= \begin{bmatrix} 3/4 & 1/2\sqrt{2} & -1/4 \\ 1/2\sqrt{2} & 1/2 & 1/2\sqrt{2} \\ -1/4 & 1/2\sqrt{2} & 3/4 \end{bmatrix}$$

Thus, $q_1 = q_3 = 3/2$; $q_2 = 1$; etc.

You should be able to carry out the multiplication to obtain

$$D_{benzene} = \begin{bmatrix} 1/2 & 1/3 & 0 & -1/6 & 0 & 1/3 \\ 1/3 & 1/2 & 1/3 & 0 & -1/6 & 0 \\ 0 & 1/3 & 1/2 & 1/3 & 0 & -1/6 \\ -1/6 & 0 & 1/3 & 1/2 & 1/3 & 0 \\ 0 & -1/6 & 0 & 1/3 & 1/2 & 1/3 \\ 1/3 & 0 & -1/6 & 0 & 1/3 & 1/2 \end{bmatrix}$$

The important point of all of this is that the density matrix contains all of the information about electron densities in the molecule, whereas the individual molecular orbitals are not nearly so unique. This is true not only for the electron densities, but also for the expectation value of the energy of the system. Note that Eq. (5-49) contains density matrix elements, not orbital coefficients. The distinction comes because the density matrix is invariant with respect to any unitary transformation of the occupied orbitals. That is, any set of occupied orbitals, C'', obtained by right multiplying C' by a unitary matrix

$$C'' = C'U \tag{5-72}$$

will result in the same density matrix:

$$C''C''* = C'UU*C'* = C'C'* = D \tag{5-73}$$

Therefore, C'' will give the same electron density function and the same expectation value for the energy of the system. Thus, C'' is just as "good" a set of orbitals as C'.

PROBLEMS

1. Note the similarity of the carbonate system

$$\begin{array}{ccc} {}^1O & & O^2 \\ & \diagdown \ \diagup & \\ & C^4 & \\ & | & \\ & O^3 & \end{array}$$

to the trimethylenemethane system on p. 134.
    a. Suggest a reason for the stability of the carbonate ion.
    b. In order to account for the greater nuclear charge of oxygen over carbon, set the value

$$<\gamma_1|F|\gamma_1> = <\gamma_2|F|\gamma_2> = <\gamma_3|F|\gamma_3> = \alpha + \beta$$

Then solve the Hückel problem for the carbonate system. Sketch the MO's and calculate the center populations.

HINT: Use the first three rows and columns of the U matrix for the all-carbon system to set up a unitary matrix with $U_{i4} = U_{4i} = 0$, except $U_{44} = 1$. Use this matrix to transform the F matrix, then note that the 2 x 2 matrix

$$\begin{bmatrix} 1 & \sqrt{3} \\ \sqrt{3} & 0 \end{bmatrix}$$

can be diagonalized with the matrix

$$U = \begin{bmatrix} 0.60 & 0.80 \\ -0.80 & 0.60 \end{bmatrix}$$

2. Calculate the "resonance stabilization" energies of each of the systems treated on p. 132-134.

3. Show that the trace of any matrix, M, $[Tr(M) = \sum_i M_{ii}]$, is invariant under a unitary transformation. That is, show that if $U*MU = N$, then $\sum_i N_{ii} = \sum_i M_{ii}$.
   How does this concern the problem of orbital energies in the Hückel approximation?

4. Show that the function

$$G = x e^{-\alpha x^2}$$

is an eigenfunction of the total energy operator for the one-dimensional particle operated on by Hooke's law force, if $\alpha = (km)^{1/2}/2\hbar$.
   Find the constant A such that $Axe^{-\alpha x^2}$ is normalized.

5. Left-multiply the equation immediately below Eq. (5-27) by F and integrate to obtain $<E>$.
   Show that the minimization of $<E>$ with respect to α gives the result shown.

6. Show that the function

$$B = \exp [\alpha(x^2 + y^2 + z^2)^{1/2}]$$

is an eigenfunction of the Hamiltonian operator for the hydrogen atom if α is properly chosen.

REFERENCES

1. S. Glasstone, Theoretical Chemistry, D. Van Nostrand Co., Inc., Princeton, New Jersey, 1944, pp. 21ff. This is a useful source for those with no acquaintance with wavefunctions.

2. A. Streitwieser, Molecular Orbital Theory for Organic Chemists, John Wiley and Sons, Inc., New York, New York, 1961. An elementary, but fairly thorough, discussion of the applications of Hückel Pi-MO theory is presented.

3. W. J. LeNoble, Highlights of Organic Chemistry, Marcel Dekker, Inc., New York, New York, 1974, pp. 277-292. A brief discussion of the chemistry of 4n Pi-electron systems and references to the original literature is given.

4. R. B. Woodward and R. Hoffmann, Angew. Chem., Int. Ed., Engl., $\underline{8}$, 781 (1969). See also: W. J. LeNoble, Highlights of Organic Chemistry, Marcel Dekker, Inc., New York, New York, 1974, Chap. 14. This chapter presents an excellent review of this field.

5. C. H. DePuy, Accts. Chem. Res., $\underline{1}$, 33 (1968).

6. L. F. Phillips, Basic Quantum Chemistry, John Wiley and Sons, Inc., New York, New York, 1965. A good introduction to formal quantum mechanics is presented.

7. R. G. Parr, Quantum Theory of Molecular Electronic Structure, W. A. Benjamin, Inc., New York, New York, 1964. This volume gives an outstandingly clear introduction to MO theory.

8. M. J. S. Dewar, The Molecular Orbital Theory of Organic Chemistry, McGraw-Hill Book Co., Inc., New York, New York, 1969. Many of these methods are discussed in this volume. See also Ref. 7.

9. There have been recent attempts to reformulate quantum theory in terms of density matrices. For these interesting attempts and the problems encountered, see Reduced Density Matrices With Applications to Physical and Chemical Systems, (A. J. Coleman and R. M. Erdahl, eds.), Queen's Papers on Pure and Applied Mathematics, No. 11, Queen's University, Kingston, Ontario, Canada, 1968.

10. M. L. Boas, Mathematical Methods in the Physical Sciences, John Wiley and Sons, Inc., New York, New York, 1966, pp. 142ff.

Chapter 6

CARBONIUM ION, CARBANION, ACID, AND
BASE EQUILIBRIA

6-1. INTRODUCTION

In previous chapters we have discussed the fact that many organic reaction mechanisms consist of several elementary steps. These steps frequently involve reactions of intermediates which can participate in simple acid–base reactions. Carbonium ions may react with water as Lewis acids:

$$R^+ + H_2O \underset{}{\overset{K_R}{\rightleftharpoons}} ROH + H^+ \qquad (6\text{-}1)$$

or, if β-protons are present, as Brϕnsted acids:

$$\underset{H}{\overset{\diagdown}{\diagup}}C - \overset{+}{C}\underset{\diagdown}{\overset{\diagup}{}} \underset{}{\overset{K_R'}{\rightleftharpoons}} \underset{\diagup}{\overset{\diagdown}{}}C = C\underset{\diagdown}{\overset{\diagup}{}} + H^+ \qquad (6\text{-}2)$$

Many reactions of carbonyl compounds, alcohols, and amines proceed through intermediates which are the Brϕnsted conjugate acids of these reactants:

$$BH^+ \underset{}{\overset{K_0}{\rightleftharpoons}} B + H^+ \qquad (6\text{-}3)$$

Carbanions are simply the Brϕnsted conjugate bases of carbon compounds:

$$R_3C\text{-}H \underset{}{\overset{K_-}{\rightleftharpoons}} R_3C^- + H^+ \qquad (6\text{-}4)$$

It is obviously of considerable interest to obtain quantitative data concerning the acid–base equilibria of these intermediates, and to find conditions under which the species can be prepared and directly observed. Unfortunately, very few acids and bases can be directly observed in aqueous solution because of the fact that water itself is both a moderately good acid and a moderately good base. The practical range of pK's which can be

161

measured in aqueous solution is limited to ca. 1-13; this is an extremely
narrow range, as we shall see in the ensuing discussion.

6-2. ACIDITIES IN AQUEOUS SOLUTION

Largely because of the fact that pH measurements [1] in aqueous solutions
were developed long before widespread interest in nonaqueous solvents, but
also because of the ubiquity of water as a solvent, measurements of acid-
base equilibria are usually referred to dilute aqueous solution as a standard
state. In Eqs. (6-5)-(6-8)

$$\frac{[ROH]}{[R+]} = K_R \frac{a_{H_2O}\gamma_{R^+}}{a_{H^+}\gamma_{ROH}} = \frac{K_R}{h_R} \tag{6-5}$$

$$\frac{[Alkene]}{[R^+]} = K'_R \frac{\gamma_{R^+}}{a_{H^+}\gamma_{alkene}} = \frac{K'_R}{h'_R} \tag{6-6}$$

$$\frac{[B]}{[BH^+]} = K_0 \frac{\gamma_{BH^+}}{a_{H^+}\gamma_B} = \frac{K_0}{h_0} \tag{6-7}$$

$$\frac{[R_3C^-]}{[R_3CH]} = K_- \frac{\gamma_{R_3CH}}{a_{H^+}\gamma_{R_3C^-}} = \frac{K_-}{h_-} \tag{6-8}$$

the activity coefficients, γ, and the activity of water, a_{H_2O}, are thus defined as unity for dilute aqueous solution. The quantity h defined by the last
equality in each of these equations is then equal to the activity of the proton,
a_{H^+}, in dilute aqueous solution.

Equations (6-5)-(6-8) are in the form most conveniently used for the
measurements of equilibrium constants. Generally, the concentration
ratio on the left-hand side of the equation is determined, frequently by
spectrophotometric methods, the pH of the solution is measured by either
indicators or electrode systems, and the activity coefficients are handled
either by extrapolation of a series of measurements to infinite dilution, or
by application of the Debye-Hückel equation.

Table 6-1 lists some equilibrium constants for reactions of carbonium
ions and carbanions in aqueous solution which were obtained by the method
outlined above.

All of the carbon acids listed in Table 6-1 have strong electron-
withdrawing groups attached directly to the acidic carbon. For most of
these compounds there is considerable evidence that the negative charge
resides largely on an electronegative atom other than carbon. For example, the nitro compound anions have structures quite close to that expected
for the classical structure:

TABLE 6-1

Equilibrium Constants in Aqueous Solution [2-7]

Carbonium ion	pK_R
Substituted triphenylmethyl cations	
p,p',p''-tris(dimethylamino)	9.36
p,p'-bis(dimethylamino)	6.85
p,p'-bis(dimethylamino)-p''-CF_3	6.13
p,p'-bis(dimethylamino)-p''-NO_2	5.51
p,p'-bis(methoxy)-p''-dimethylamino	5.75
p-methoxy-p'-dimethylamino	4.86
p-methyl-p'-dimethylamino	4.40
p-dimethylamino	3.88
tris(2,4-dimethoxy)	3.28
p,p',p''-trimethoxy	0.82
p-Dimethylaminophenyl tropylium ion	7.35
p-Methoxyphenyl tropylium ion	5.75
Phenyl tropylium ion	4.85
p-Chlorophenyl tropylium ion	4.55
Tropylium ion	4.77
1,2,3-Tri-i-propyl cyclopropenium ion	7.2

Acid	pK_-
Trinitromethane	<0
Dinitromethane	3.6
Nitromethane	10.2
Acetylacetone	9.0
Triacetylmethane	5.8
Nitroethane	8.6
Malononitrile	11.2
Cyanoform	<0
Acetic acid	4.8
Phenol	9.9
Pentacyanocyclopentadiene	<<0

$$\overset{R}{\underset{R}{\diagdown}}C = N+\overset{\diagup O^-}{\underset{\diagdown O^-}{}}$$

Similarly, for the ketones the negative charge resides largely on the oxygen of the enolate anion. We shall see in our discussion of the rates of reactions of carbanions in a later chapter that this difference in structure of the acid and its conjugate base is responsible for at least a part of the activation energies for proton transfer reactions of these compounds.

The carbonium ions listed in Table 6-1 are also unusually stable because of charge delocalization. Note particularly that the tropylium and cyclo-propenium ions, which fit the Hückel 4n + 2 rule, have stabilities compar-able to triarylmethyl cations having electron-donating groups in the p-positions.

Unfortunately, Table 6-1 nearly exhausts the list of carbonium ions and carbanions which may be studied in aqueous solution.

6-3. ACIDITY FUNCTIONS [8]

When we leave dilute aqueous solutions, the h's defined in Eqs. (6-5)-(6-8) are no longer equal to proton activity, but they still determine the concen-tration ratio of acid to conjugate base in the same sense that proton activity does in dilute aqueous solution. Thus, for each particular type of base, h is a measure of the ability of the solution to donate a proton to that base to form the conjugate acid.

The question of whether h will be a "useful" function will depend on whether it has the same value for some predictable range of structures of the bases. That is, if we could predict that some series of bases B would all have the same values of γ_{BH^+}/γ_B in a range of solvents, then the func-tion h_0 would be the same for all of these bases, and would, for example, be useful in determining K_0 values.

There are two problems involved, then, in the quantities h. First, can we make reasonable predictions about which bases are likely to have the same activity coefficient ratios, and second, if we can, how can one es-tablish a scale of h values for various solvents of interest?

The behavior of activity coefficients was discussed in Chapter 2, and it was pointed out that they are determined in dilute solutions by interactions of the solute species with solvent. Hydrogen bonding, either from solute to solvent or from solvent to solute, was stressed to be one of the more im-portant types of interactions which affect activity coefficient behavior on change of solvent. Other interactions, such as dipole-dipole, charge-dipole, and dispersion forces, were also mentioned. For the present pur-poses, it is important to realize that many of these interactions will be nearly identical for both the acid and conjugate base forms of the solutes under consideration. For example, the activity coefficient ratio, γ_{BH^+}/γ_B, for the protonation of m-nitroaniline will depend primarily on the change of the NH_2 group to the NH_3^+ group. Solvent interactions with the nitro group, which almost certainly affect both γ_{BH^+} and γ_B, should be nearly the same for the base and the conjugate acid, and should nearly cancel in the ratio of the activity coefficients.

From this type of reasoning, we might expect that the ratio $\gamma_{R^+}/\gamma_{ROH}$ will be nearly constant for a wide range of alcohols, but will be different from $\gamma_{R^+}/\gamma_{alkene}$, since the OH group will almost certainly have different interactions with solvent than does the double bond of the alkene. Similarly, we expect that γ_{BH^+}/γ_B will not be the same for primary and secondary amines, and that $\gamma_{R_3CH}/\gamma_{R_3C^-}$ will be different for the ionization of ketones from that for the ionization of hydrocarbons such as triphenylmethane.

Although care, experience, and intelligence are required in choosing compounds for which the same h function will apply, there is enough generality and predictability that the functions are indeed useful.

The second problem, that of establishing h scales, is nearly a circular problem. If we knew the K's for a series of similar compounds, we could establish an h scale from measurements of concentration ratios and applications of Eq. (6-5), (6-6), (6-7), or (6-8); and, if we knew an h scale for solvents in which we could measure concentration ratios, we could use the equations to evaluate K's. The problem can be solved, however, by a "bootstrap" method called the "overlapping indicator technique." The method involves finding a series of compounds, for all of which we expect the same h function to apply, and which have closely spaced K values.

As an example, suppose we have a series of p-, p'-, and p''-substituted triarylcarbinols, Ar_3COH, with substituents chosen such that the K's are expected to vary regularly with substitution. It is found that the concentration ratio, $[Ar_3C^+]/[Ar_3COH]$, for trianisylcarbinol can be measured in aqueous solution buffered at pH = 2. At the low ionic strength in this solution, the activity coefficients in Eq. (6-5) can be set equal to unity for the uncharged species, and can be estimated from Debye-Hückel theory for the charged species. Thus, K_R can be directly evaluated for the trianisyl-methyl cation, and pK_R is found to be 0.82.

If the trianisylcarbinol is placed in 5% by weight sulfuric acid in water, the concentration ratio, $[Ar_3C^+]/[Ar_3COH]$, is found to be 7.73. Since we know K_R, we may solve Eq. (6-5) to obtain $h_R = 1.17$ for the 5% H_2SO_4 solution.

If p,p'-dimethoxytriphenyl carbinol is dissolved in the same 5% by weight H_2SO_4 solution, the concentration ratio, $[Ar_3C^+]/[Ar_3COH]$, is found to be 0.068. Substitution of this value into Eq. (6-5) along with the h_R value determined above, gives $K_R = 17.4$, or $pK_R = -1.24$, for this cation. This carbinol could now be dissolved in 20% by weight sulfuric acid, the concentration ratio measured, and the h_R value for the 20% H_2SO_4 thereby determined.

This procedure can be repeated as long as we can find two carbonium ions, both of whose concentration ratios can be measured in the same solution. In actual practice the concentration ratios can generally be measured accurately only if $0.01 < (Ar_3C^+)/(Ar_3COH) < 100$. Thus, the concentration ratio for a particular cation can be measured roughly over a range of four units of $\log h_R$.

In general, if we choose a set of compounds whose pK's are spaced about one unit apart, we can measure concentration ratios for any two compounds over three units of $\log h$. If our hypothesis is true that $\gamma_{R^+}/\gamma_{ROH}$ is the same for all of the compounds studied, then measurements of the concentration ratios for any pair of compounds in any solution within this "overlap" region must give consistent results when the values are substituted into Eq. (6-5). That is, the equation

$$\frac{K_{R_1}}{K_{R_2}} = \frac{[R_1OH]\ [R_2^+]}{[R_2OH]\ [R_1^+]} \tag{6-9}$$

must be satisfied for all solutions in which both concentration ratios can be measured.

In Problem 1 at the end of this chapter, some typical data are given for the overlap technique applied to the protonation of primary anilinium ions in sulfuric acid solutions. The data show the expected consistency for the equation analogous to Eq. (6-9). It should be obvious at this point that the overlap can be used to derive any of the h scales defined in Eqs. (6-5)-(6-8). For convenience of notation, acidity functions H, analogous to pH, are defined as $-\log h$.

An H_R acidity function has been established for many aqueous-acid mixtures, such as H_2SO_4-H_2O, $HClO_4$-H_2O, HCl-H_2O, by the use of the triarylcarbinols discussed in our example above.

An H_R' function has been established according to Eq. (6-6) by the use of substituted azulenes, which undergo the following protonation reaction:

Studies have been done in sulfuric acid and perchloric acid solutions only.

There are several H_0 functions which have been established in a variety of aqueous and nonaqueous acid solutions. The protonation of amides, indoles, primary anilines, and tertiary anilines each generate a distinct H_0 scale. The scale based on primary anilines is generally denoted H_0', and that based on tertiary anilines is denoted H_0''.

Some of these acidity functions for sulfuric acid solutions are listed in Table 6-2.

The function H_R differs from the other acidity functions in that the activity of water appears in its definition. The activity of water in sulfuric acid solutions can be determined by measurements of the vapor pressure of water in equilibrium with the solutions. The last column of Table 6-1 reports the values of $H_R - \log a_{H_2O}$, which would be equal to the other acidity functions if the activity coefficient ratios were equal.

TABLE 6-2

Acidity Functions for H_2O-H_2SO_4 Solutions [8]

Wt percent H_2SO_4	$-H_R$	$-H_R'$	$-H_0'''$	$-H_0'$	$-(H_R - \log a_{H_2O})$
5	0.07	0.27	-0.02	–	0.10
10	0.72	0.51	0.53	0.31	0.70
20	1.92	1.50	1.47	1.01	1.86
30	3.22	2.55	2.44	1.72	3.10
40	4.80	3.85	3.46	2.41	4.55
50	6.60	5.53	4.54	3.38	6.15
60	8.92	7.34	5.91	4.46	8.13
70	11.52	9.40	7.65	5.80	10.18
80	14.12	–	9.44	7.34	11.84
85	15.42	–	10.30	8.14	12.54
90	16.72	–	11.14	8.92	13.23
92	17.24	–	–	9.29	–
94	17.78	–	–	9.68	–
95	18.08	–	11.89	–	–
96	18.45	–	–	10.03	–
98	19.64	–	–	10.41	–
99.5	–	–	–	11.30	–

The qualitative behaviors of the various acidity functions can be understood in terms of the effect of decreasing water activity with increasing
percent H_2SO_4 on the activity coefficients of the proton and of the conjugate
acid involved in the equilibrium. As water activity decreases, the hydration of the proton (i.e., H_3O^+, $H_9O_4^+$, etc.) becomes more difficult, which
raises the activity coefficient of the proton (i.e., it becomes less "stable").
At least part of this effect is balanced if the conjugate acid involved also
requires water of hydration for its stability. The primary anilinium ions
have three acidic protons which can hydrogen bond to water molecules providing stabilization for this conjugate acid. The tertiary anilinium ions
only have one acidic hydrogen, and are, therefore, less stabilized by hydration. The function H_0''', therefore, increases more rapidly than H_0' as
water activity decreases in the solutions.

There is considerable evidence that carbonium ions, even those with β-
protons, are extremely poor hydrogen bond donors [9]. The activity coefficients of the carbonium ions, therefore, are even less sensitive to water
activity than the tertiary anilinium ions, and both H_R' and H_R - log a_{H_2O}
increase more rapidly than either of the H_0 functions. The difference between the two carbonium ion acidity functions (i.e., H_R' and H_R - log a_{H_2O})
is probably caused by the different hydrogen bonding of solvent to the OH
group and to the olefinic bond of the conjugate bases.

The acidity function H_- based on reaction 4 type acids is most useful in
highly basic solvents rather than in the highly acidic solvents of interest
for the other acidity functions. Such H_- scales have been established for a
number of solutions, such as aqueous sodium hydroxide, sodium hydroxide
in water-alcohol mixtures, and tetramethylammonium hydroxide in water-
dimethylsulfoxide solutions, by the use of carbon acids, and by the use of
amines ionizing in the following sense:

$$ArNH_2 \rightleftharpoons ArNH^- + H^+$$

We shall discuss some of these functions more thoroughly in our later
consideration of carbanion reactions. Simply to illustrate the range of
acidity function values available, the H_- values for 0.011 M $(CH_3)_4N^+OH^-$
in water-dimethylsulfoxide solvents determined by the use of some amines
are listed in Table 6-3.

Again, simply to illustrate the range of pK values which can be determined by the use of acidity functions, the pK values of a number of compounds are reported in Table 6-4.

It is particularly striking to note that the strongest acid listed in Table
6-4, the tris(p-nitrophenyl)methyl cation, would have a concentration ratio
$[Ar_3C^+]/[Ar_3COH]$ of $10^{-23.7}$ in neutral aqueous solution (i.e., one molecule of the carbonium ion per mole of the alcohol), and that the weakest
acid, triphenylmethane, would have a concentration ratio $[Ar_3CH]/[Ar_3C^-]$
of $10^{+21.2}$ in neutral aqueous solution.

TABLE 6-3

H_- Acidity Function For 0.011 M $(CH_3)_4N^+OH^-$ in
DMSO-H_2O Solutions [8]

Mole percent DMSO	H_-	Mole percent DMSO	H_-
10.32	13.17	76.12	20.14
15.20	13.88	83.14	20.97
23.57	14.86	88.79	21.61
30.11	15.54	92.47	22.45
36.79	16.17	95.77	23.32
43.27	16.83	97.13	23.88
52.55	17.73	98.29	24.50
58.56	18.34	98.71	24.84
69.09	19.41	99.14	25.30
		99.59	26.19

6-4. ACIDITIES OF HYDROCARBONS

Since compounds such as the simple alkyllithiums have been known for
some time, it was once hoped that direct measurements of the exchange
equilibria, i.e.,

R-Li + R'-H $\rightleftharpoons$ R-H + R'-Li

would give information directly related to carbanion stabilities. It has
turned out, however, that this method has at least two nearly insurmount-
able difficulties. First, in those solvents in which alkyllithium compounds
are stable, aggregates (commonly tetramers and hexamers) are the stable
species in solution [11]. It appears that steric effects within the aggregates are more important in determining the position of the equilibria than
are the "intrinsic" stabilities of the alkyllithiums, Second, the carbon-
lithium bond in simple alkyllithiums is now known to be highly covalent.
Thus, it is unlikely that the equilibrium constants for exchange of mono-
mers would be related to carbanion stabilities even if they could be measured.

TABLE 6-4

pK Values Determined by Acidity Function Methods [8,10]

Acid	pK	Acid	pK
Substituted anilines		Substituted triphenylmethyl	
3-Cl	25.63	cations	
3-CN	24.64	p,p,p''-OCH$_3$	0.82
3,5-Cl$_2$	23.59	p,p'-OCH$_3$	-1.24
4-CN	22.68	p-OCH$_3$	-3.40
2,3,5,6-Cl$_4$	19.22	p,p',p''-CH$_3$	-3.56
4-Cl. 2-NO$_2$	17.08	p-CH$_3$	-5.41
N-ph	22.44	m,m',m''-CH$_3$	-6.35
N-Ph, 3,4-(NO$_2$)$_2$	14.66	H	-6.61
N-Ph, 2,4-(NO$_2$)$_2$	13.84	p,p',p''-Cl	-7.74
		p-NO$_2$	-9.15
Carbon acids		p,p',p''-NO$_2$	-16.27
Malononitrile	11.14		
9-Cyanofluorene	11.41	Other acids	
Cyanoform	-4.85	Azulenium ion	-1.7
Fluorene	21.0	p,p'-Dimethoxybenzhydryl	
Triphenylmethane	28.2	cation	-5.71
9-CO$_2$CH$_3$ fluorene	12.9	2,4,6-Trinitrodimethyl-	
9-Phenylfluorene	18.6	anilinium	-6.55

Many carbon acids considerably weaker than these listed in Table 6-1, but considerably stronger than simple aliphatic hydrocarbons, can be ionized in the presence of bases in nonaqueous solutions. The fluorenyl anion, for example, can be prepared in ether solution by the reaction of fluorene with potassium t-butoxide:

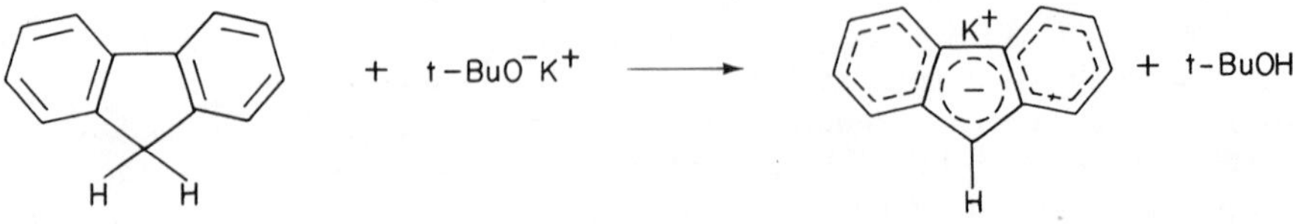

In such solvents, however, the carbanions exist only as ion pairs or higher ionic aggregates. Although a great deal of information has been accumulated on equilibrium constants for exchange reactions of carbanion ion pairs in solvents of low dielectric constant [7,12], it is doubtful that these can be related directly to the stabilities of the "free" carbanions.

In recent years it has been found that dimethylsulfoxide is a particularly favorable solvent for the study of moderately stable carbanions [12,13]. The solvent has a relatively high dielectric constant ($\epsilon = 48$ at 25 °C), and also strongly solvates alkali metal cations. Moreover, the conjugate base of the solvent, the dimsyl anion ($CH_3SOCH_2^-$), can be cleanly prepared in the solvent and is a strong enough base to quantitatively remove a proton from acids stronger than triphenylmethane. Thus, it is possible to prepare stable solutions of many carbanions, and by working in dilute solutions, ion pairing problems can be avoided. A pH scale for the solvent has been established for pH lower than ca. 12, and measurements of the exchange reactions, i.e.,

$$R^- + R'\text{-}H \xrightleftharpoons{K_{ex}} R\text{-}H + R'^-$$

$$K_{ex} = \frac{[RH]\,[R'^-]}{[R^-]\,[R'\text{-}H]}\, \frac{\gamma_{RH}\gamma_{R'-}}{\gamma_{R^-}\gamma_{R'H}} \tag{6-10}$$

can be made to obtain pK values relative to a standard state in dilute dimethylsulfoxide solutions [7,12,13] over a wide range. Some of the values obtained are shown in Table 6-5, along with pK values in water and in methanol solutions for comparison.

It is quite clear from the data in Table 6-5 that the relative acidities determined in dimethylsulfoxide solution are quite different in many cases from those determined in the hydroxylic solvents. For example, the pK of malononitrile is 6.2 units lower than that of nitromethane in dimethylsulfoxide solution, but is one unit higher than that of nitromethane in water. To state the facts in a different way, the activity coefficient ratio in Eq. (6-10) has a value of $10^{7.2}$ in dimethylsulfoxide solution relative to a standard state in aqueous solution for the exchange of malononitrile anion with nitromethane. Even larger values for this activity coefficient ratio are found for exchange of alkoxide ions with the hydrocarbons. The ratio has a value of 10^{18} for the exchange of 9-cyanofluorenide ion with methanol in dimethylsulfoxide solution relative to a standard state in methanol solution.

These data show clearly that the H_- acidity function applied to hydrocarbons is, at best, limited to extremely similar structures, and that the pK values in Table 6-4, supposedly referred to a standard state in aqueous solution, are of doubtful significance. Qualitatively, it appears that hydrogen bonding is the primary factor responsible for most of the large values of the activity coefficient ratios in Eq. (6-10). Those anions, such as RO^- and $RC{\equiv}C^-$, which have a localized lone pair of electrons can accept strong hydrogen bonds from hydroxylic solvents. Those anions having only delocalized lone pairs cannot accept strong hydrogen bonds.

TABLE 6-5

pK's in Various Solvents [7, 12, 13]

Acid	pK		
	H_2O	MeOH	DMSO
Benzoic acid	4.2	9.4	10.8
Acetic acid	4.8	9.7	11.9
Nitromethane	10.2	–	17.2
Nitroethane	8.6	–	16.7
Malononitrile	11.1	–	11.0
Acetylacetone	9.0	–	13.5
$(C_2H_5)_3NH^+$	10.9	10.9	9.0
9-Cyanofluorene	–	14.2	8.3
9-Carbomethoxyfluorene	–	–	10.3
Fluoradene	–	–	10.5
p-Nitrophenol	7.1	11.2	10.4
9-Phenylfluorene	–	–	17.9
9-Methylfluorene	–	–	22.3
Fluorene	–	–	22.6
Methanol	–	16.9	29.1
Ethanol	–	–	29.5
t-Butanol	–	–	31.3
Phenylacetylene	–	–	28.8
Acetone	–	–	26.5
Acetonitrile	–	–	31.2
Triphenylmethane	–	–	30.6
Acetophenone	–	–	24.7

Because of this rather obvious importance of hydrogen bonding, there has been some thought that relative acidities measured in nonhydroxylic solvents, such as dimethylsulfoxide, are more "intrinsic" than those measured in the more common hydroxylic solvents. It has been supposed that specific types of solvent-solute interactions in the nonhydroxylic solvents are unlikely to change the orders of acidity from those intrinsically associated with the solute species. These ideas, however, are almost certainly not correct.

The use of recently developed instrumentation has allowed the measurement of relative acidities of many compounds in the gas phase [14-16]. The observed order of intrinsic acidities, for example, t-BuOH > EtOH > Toluene > MeOH > H_2O, is strikingly different from that observed in any solvent thus far studied. At the present time it is clear that polarizability is an extremely important factor in determining the stabilities of ions in the gas phase. Any matter, such as hydrocarbon groups, in the vicinity of a charge is polarized by the charge, resulting in a lowering of the energy of the system from what it would be in the absence of the polarizable matter. In the alcohols, for example, the order of gas phase acidities depends only on the size of the alkyl groups in close proximity to the negatively charged oxygen. Similarly, pyridine is more basic in the gas phase than is ammonia simply because pyridinium ion has the charge in the presence of more polarizable matter than does ammonium ion. As expected from resonance effects, piperidine is a stronger base than pyridine both in the gas phase and in solutions. In this case, the "normal" structural effects, which we have discussed in earlier chapters, are seen to operate in the gas phase if they are not swamped out by the polarizability effects.

It is clear that polarizability of the molecule itself is less important in solution than in the gas phase. This probably results from the fact that solvent molecules in the vicinity of ions serve as polarizable matter just as well as do parts of the solute species. It is not clear, however, what type of solvent-solute interactions, even in nonhydroxylic solvents, are responsible for the reversal of "intrinsic" orders of acidity. As a specific example, it is not at all clear why methanol is a stronger acid than either toluene or ethanol in dimethylsulfoxide solution while the reverse is true in the gas phase.

6-5. ION PAIRS OF CARBANIONS [17]

We have already mentioned that in solvents of low dielectric constant, carbanion salts generally exist as ion pairs or as higher ionic aggregates. In solvents of moderate dielectric constant, such as ethers, pyridine, or cyclohexylamine, ion pairs are the predominant species at concentrations roughly in the range of $10^{-6}-10^{-2}$ M. Some ion pair dissociation constants for fluorenyl salts in tetrahydrofuran solution are shown in Table 6-6. These show clearly that free ions are not present to any appreciable extent at practical concentrations of the carbanions.

The ion pair dissociation constants are usually measured by the use of conductivity, but UV and visible spectra also indicate the presence of ion pairs since the wavelength of maximum absorption of the fluorenyl anion depends on the identity of the cation present. For the Li^+, Na^+, K^+, Cs^+, and $(n\text{-}Bu)_4N^+$ salts of the fluorenyl anion in THF solution, λ_{max} is 349, 356, 362, 364, and 368 nm, respectively, at 25°C.

It is also observed that the spectrum of fluorenyl salts in THF is temperature dependent. The UV and visible spectra of fluorenyl sodium in THF at 25°C and at -50°C are shown in Fig. 6-1. At temperatures between these two, the spectrum is intermediate between those shown. The constancy of absorption at those wavelengths where the spectra in Fig. 6-1 cross (isosbestic points) indicate that there are two species in solution whose proportions depend on temperature. At constant temperature the spectrum is unchanged by addition of $Na^+BF_4^-$ or by change in the concentration of the fluorenyl salt. These observations rule out the possibility that the two species represent two different degrees of ionic aggregation.

The above data, now supplemented by a large number of observations on similar types of systems, are indicative of two different types of ion pairs, believed to be the intimate and solvent-separated ion pairs analogous to the species which we discussed in connection with carbonium ion reactions. The species absorbing at the longer wavelength, and predominating at the lower temperature, is the solvent-separated ion pair. That one absorbing at the shorter wavelengths and predominating at the higher temperature is the intimate ion pair.

TABLE 6-6

Ion Pair Dissociation Constants of Fluorenyl Salts in
THF Solution [18]

T (°C)	K_d (M)	
	Cs^+Fl^-	Na^+Fl^-
25	1.4×10^{-8}	6.2×10^{-7}
10	1.8×10^{-8}	1.3×10^{-6}
0	2.1×10^{-8}	2.1×10^{-6}
-20	2.8×10^{-8}	6.7×10^{-6}

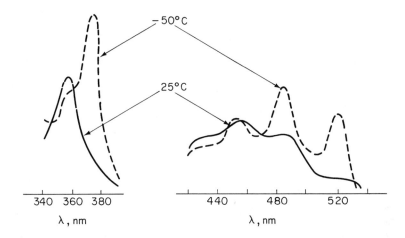

Figure 6-1. The UV and visible spectra of fluorenyl sodium in THF.

The ratio of intimate to solvent-separated ion pairs depends on cation and solvent as well as upon temperature. Some illustrative data are shown in Table 6-7.

6-6. DIRECT OBSERVATIONS OF SOME SIMPLE CARBONIUM IONS [19]

Although the acidity functions have been very useful in allowing the direct observation of many "unstable" carbonium ions and in providing quantitative measures of relative stabilities of such species, the solvents used have not been suitable for the preparation of simple alkyl carbonium ions. The H_R function for many of the solvent systems is probably low enough that the equilibrium of Eq. (6-1) is driven far to the left, but apparently the H_R' function is not low enough to drive the equilibrium of Eq. (6-2) far enough to the left. The result is that high enough concentrations of alkene are formed from alkyl carbonium ions that cationic polymerization takes place, usually resulting in the formation of conjugatively unsaturated carbonium ions.

The use of "super-acid" solvents, consisting of mixtures of HSO_3F and SbF_5, usually with some SO_2 added to improve fluidity, has allowed the preparation of a wide variety of alkyl carbonium ions which can be studied in these solutions, although quantitative acidity functions cannot be developed. Several spectroscopic techniques, NMR, UV, Raman, and photoelectron, have been applied in establishing structures of the cations and in studying the reactions of these ions. Since there is no way to go in a

TABLE 6-7

Intimate and Solvent-Separated Ion Pairs of
Fluorenyl Salts at 25 °C [17]

M^+	Solvent	λ (M^+Fl^-)	λ ($M^+//Fl^-$)	Percent ($M^+//Fl^-$)
Li^+	Dioxane	346	–	0
Na^+	Dioxane	353	–	0
Li^+	Toluene	348	–	0
Li^+	2-Methyltetrahydrofuran	347	373	25
Na^+	2-Methyltetrahydrofuran	355	373	0
Li^+	Tetrahydrofuran	349	373	80
Na^+	Tetrahydrofuran	356	373	5
Li^+	Pyridine	–	373	100
Na^+	Pyridine	–	373	100
K^+	Tetrahydrofuran	362	–	0
Cs^+	Tetrahydrofuran	364	–	0
$(n-Bu)_4N^+$	Tetrahydrofuran	368	–	0

continuous manner from aqueous solution to these super-acid solutions, it
has not been possible to establish acidity functions. From various semiquantitative observations, however, the H_R' function has been estimated to
have a value of ca. -18 in 1 M SbF_5 in HSO_3F. There have been prelimin-
ary reports of attempts to establish relative stabilities of the carbonium
ions by measurements of the heats of formation of the ions from the alcohols
in these solvents [19].

The carbonium ions can be prepared from a variety of precursors in
these solvents. Alkyl halides, alcohols, and olefins all react to form the
corresponding ions. Acyl halides form the corresponding $R-C=O^+$ cations,
which can also be formed from the reaction of R^+ with CO in these solutions.
It has not, however, been possible to form primary carbonium ions. At
low temperatures, the primary alcohols exist as the protonated alcohol, and
when the temperature is raised, carbonium ions resulting from rearrangements are formed. A variety of data indicate that the primary cations are
at least 14 kcal/mole less stable than secondary cations, which, in turn, are
about 10-13 kcal/mole less stable than tertiary cations.

Among the most striking observations to come from these studies in super-acids are the extremely fast carbonium ion rearrangements, and the ubiquity of protonated cyclopropanes [20] as either intermediates or transition states in the rearrangements. The protonated cyclopropanes are presently believed to have a structure which involves a penta-coordinated carbon atom:

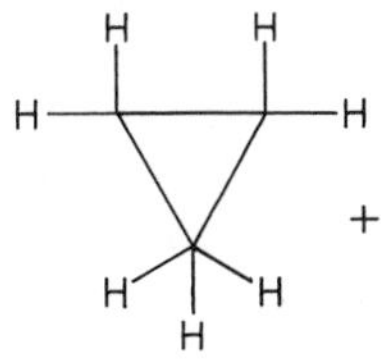

The CH_5^+ ion is a known species in the gas phase, and represents the parent penta-coordinated carbon. Although this parent ion is not observed in super-acid solutions, there is considerable evidence, based mostly on H–D exchange of CD_4, that it can be at least a steady state intermediate in such solutions. Cyclopropane itself undergoes rapid proton exchange reactions even in sulfuric acid solutions.

The 2-propyl cation has been studied extensively by the use of both proton and ^{13}C NMR. At room temperature, or slightly higher, all three carbons and all seven hydrogens are rapidly interchanged by processes believed to involve both 1,2-hydrogen shifts and protonated cyclopropane intermediates [20]:

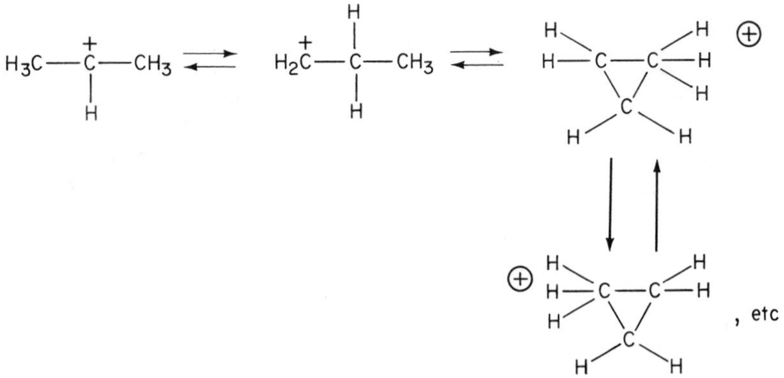

Similarly, the cyclopropylcarbinyl cation undergoes rapid scrambling reactions believed to proceed through a bicyclobutyl intermediate:

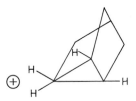

In both of the above systems, the movement of the proton around the cyclopropane ring is indicated to be extremely rapid, and the protonated cyclopropanes are not observable. That is, the classical structures of the cations are more stable than the penta-coordinated structures.

In the case of the 2-norbornyl cation, however, there is fairly convincing evidence from Raman and photoelectron spectroscopy that the stable structure is that with a protonated cyclopropane ring [21]:

At temperatures above 13 °C, this cation undergoes rearrangements which scramble all protons rapidly. At temperatures below 13 °C, the NMR spectrum shows three types of protons, and only when the temperature is lowered to less than –120 °C does one see the five types of protons expected for the cation. The reaction scheme postulated for these observations is the following:

The studies in super-acid solutions, then, have furnished strong evidence confirming and extending the conclusions about neighboring group participation in solvolysis reactions which we discussed in Chapter 3. In addition to the cases above which relate to the participation by saturated carbon, stable cationic species corresponding to the intermediates postulated for halogen, oxygen, and double bond participation have been directly observed.

PROBLEMS

1. The pK of 2,5-dichloro, 4-nitroanilinium ion is -1.78. The following table contains the values of log [B]/[BH$^+$] found for various substituted anilinium ions in aqueous sulfuric acid solutions. Calculate the pK of each of the anilinium ions, and the H_0 function for each of the solutions.

	Log [B]/[BH$^+$]					
	Substituents on Anilinium Ion					
Wt percent H_2SO_4	2,5-Cl$_2$- 4-NO$_2$	2-Cl- 6-NO$_2$	2,6-Cl$_2$- 4-NO$_2$	2,4- (NO$_2$)$_2$	2,6- (NO$_2$)$_2$	2,6-(NO$_2$)$_2$- 4-Cl
19.38	0.80					
21.98	0.64					
29.35	0.13	0.74				
37.19	-0.44	0.17	1.10			
41.63	-0.77	-0.11	0.72			
43.95	-0.94	-0.31	0.49			
45.72		-0.47	0.36			
52.21		-0.98	-0.33	0.86		
57.38			-0.84	0.35		
59.77				0.10	1.21	
61.94				-0.15	0.88	
63.50				-0.50	0.61	1.21
68.65				-1.07	-0.05	0.58
74.82					-1.04	-0.49
77.50						-0.89

2. Explain why the azulenes used in establishing the H_R' function do not undergo reaction (6-1), show no tendency for cationic polymerization, and why azulene itself has a large dipole moment.
   HINT: Consider the 4n + 2 rule.

3. From the data given in Table 6-4, calculate the Hammett ρ value for the ionization of triarylmethanols.

4. Explain how you might be able to measure the equilibrium constant, by the use of acidity functions, for the following reaction:

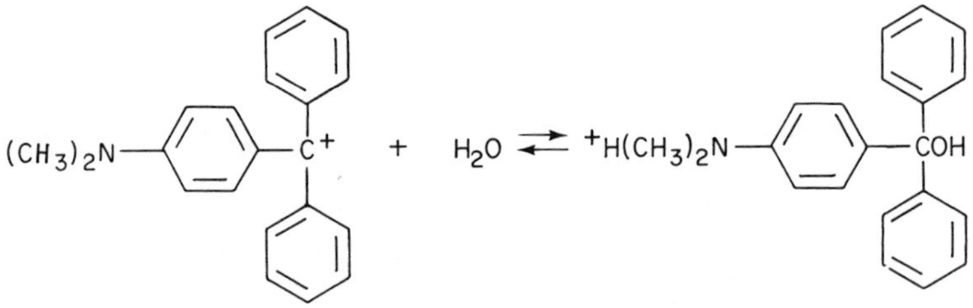

5. If a reaction is believed to proceed by the mechanism

$$B + H^+ \underset{k_{-1}}{\overset{k_1}{\rightleftharpoons}} BH^+ \overset{k_2}{\longrightarrow} \text{products}$$

with k_{-1} greater than k_2, how would the observed pseudo-first-order rate constants vary in aqueous sulfuric acid solvents?

REFERENCES

1. R. G. Bates, Determination of pH, Theory and Practice, John Wiley and Sons, Inc., New York, New York, 1964.

2. A. Streitwieser, Molecular Orbital Theory for Organic Chemists, John Wiley and Sons, Inc., New York, New York, 1961, Chap. 12.

3. D. Lloyd, Carbocyclic Non-Benzenoid Aromatic Compounds, Elsevier Publishers, Amsterdam, Netherlands, 1966.

4. C. D. Ritchie and H. Fleischhauer, J. Amer. Chem. Soc., 94, 3481 (1972).

5. C. D. Ritchie, D. J. Wright, D. Huang, and A. Kamego, J. Amer. Chem. Soc., 97, 1163 (1975).

6. D. J. Cram, Fundamentals of Carbanion Chemistry, John Wiley and Sons, Inc., New York, New York, 1965, Chaps. 1 and 2.

7. C. D. Ritchie in Solvent-Solute Interactions (J. F. Coetzee and C. D. Ritchie, eds.), Marcel Dekker, Inc., New York, New York, 1969, Chap. 4.

8. R. H. Boyd in Solvent-Solute Interactions (J. F. Coetzee and C. D. Ritchie, eds.), Marcel Dekker, Inc., New York, New York, 1969, Chap. 3.

9. J. F. Bunnett, J. Amer. Chem. Soc., 83, 4956-4976 (1961).

10. K. Bowden and R. Stewart, Tetrahedron, 21, 261 (1965). See also Ref. 8.

11. J. F. Garst in Solvent-Solute Interactions (J. F. Coetzee and C. D. Ritchie, eds.), Marcel Dekker, Inc., New York, New York, 1969, Chap. 8.

12. W. S. Matthews, J. E. Bares, J. E. Bartmess, F. G. Bordwell, F. J. Cornforth, G. E. Drucker, Z. Margolin, G. J. McCollum, and N. R. Vanier, J. Amer. Chem. Soc., July 1975 (in press).

13. F. G. Bordwell and W. S. Matthews, J. Amer. Chem. Soc., 96, 1214, 1216 (1974).

14. R. Yamdagni and P. E. Kebarle, J. Amer. Chem. Soc., 95, 4050 (1973).

15. J. I. Brauman and L. K. Blair, J. Amer. Chem. Soc., 93, 4315 (1971); 90, 6561 (1968).

16. E. M. Arnett, Accts. Chem. Res., 6, 404 (1973); see also earlier references cited therein.

17. J. Smid in Ions and Ion-Pairs in Organic Reactions (M. Szwarc, ed.), Wiley Interscience, Inc., New York, New York, 1972, Chap. 3.

18. M. Szwarc in Ions and Ion-Pairs in Organic Reactions (M. Szwarc, ed.), Wiley Interscience, Inc., New York, New York, 1972, Chap. 1.

19. G. A. Olah and P. v. R. Schleyer, Carbonium Ions, Wiley Interscience, Inc., New York, New York, 1968. This is a four volume series which covers a wide range of carbonium ion studies.

20. M. Saunders, P. Vogel, E. L. Hagen, and J. Rosenfeld, Accts. Chem. Res., 6, 53 (1973).

21. G. A. Olah, J. Amer. Chem. Soc., 95, 8698 (1973).

Chapter 7

ACID-BASE CATALYSIS

7-1. BUFFER SYSTEMS

There are a very large number of organic reactions which are catalyzed by Brønsted acids or bases. Such reactions may be classified into two very broad categories depending upon whether the proton transfer step of the reaction is rate determining or not. We have already discussed, in other connections, reactions which involve both types of catalysis. For example, the neighboring group participation of O^- or COO^- groups in solvolysis reactions involves a rapid equilibrium between the conjugate acids and bases of these groups, followed by a rate-determining step. In these cases, the proton transfer occurs in a rapid equilibrium, not in the rate-determining step. The reaction of diphenyldiazomethane with acids, which we discussed in connection with ion pair intermediates, on the other hand, involves a slow proton transfer reaction forming the benzhydryl diazonium ion which then undergoes rapid steps to form product.

The experimental distinction between these two categories is frequently based on kinetic studies in buffered solutions. Before proceeding to the discussion of kinetics of acid and base catalysis, then, it is worthwhile to review some of the concepts of buffer systems.

Essentially, a buffer system may be composed of any Brønsted acid-base pair in equilibrium:

$$BH \rightleftharpoons B + H^+ \tag{7-1}$$

(The charges are omitted for the sake of generality.) The particular acid-base pair will have associated with it an equilibrium constant K_a defined as follows:

$$K_a^{\circ} = \frac{a_B a_{H^+}}{a_{BH}} \tag{7-2}$$

In most solutions where buffer systems are used, particularly in the study of acid-base catalysis, the ionic strength of all solutions used will be kept constant to minimize variations in activity coefficients. In the following discussion, then, we will assume that activity coefficients are indeed constant in all solutions, and we shall work with the "practical" equilibrium constants defined in Eq. (7-3):

$$K_a = \frac{[B] \, [H^+]}{[BH]} = K_a^\circ \, \frac{\gamma_{BH}}{\gamma_B \gamma_{H+}} \tag{7-3}$$

Rearrangement of Eq. (7-3) gives us what might be called the "master equation" for buffer solutions:

$$[H^+] = K_a \, \frac{[BH]}{[B]} \tag{7-4}$$

The important feature of Eq. (7-4) is that the buffer ratio, $[BH]/[B]$, is the quantity that determines $[H^+]$. That is, we can vary the concentrations of both B and BH without changing $[H^+]$. We shall see that it is just as important that we may vary $[H^+]$ without changing $[B]$, or without changing $[BH]$, if we vary the other.

From a practical standpoint, it is essential to realize that the values of $[B]$ and $[BH]$ in Eq. (7-4) are not necessarily equal to the values which we would calculate from the amount of B and BH added to the solution. Suppose we weigh out a certain number of moles of B (W_B), and a certain number of moles of BH (W_{BH}) and dissolve these in a volume V of water. We define the following quantities:

$$C_B^\circ = \frac{W_B}{V}$$

$$C_{BH}^\circ = \frac{W_{BH}}{V} \tag{7-5}$$

In order to calculate the concentrations of B and BH in solution, we note that the conversion of BH into B will produce a proton, or that the conversion of B into BH will consume a proton. The only other source of protons in solution is the equilibrium

$$H_2O \rightleftharpoons H^+ + OH^- \tag{7-6}$$

with equilibrium constant $K_w = 1.0 \times 10^{-14} \, M^2$ at 25 °C. For every proton produced by reaction (7-6), there is also produced a hydroxide ion. Thus, the only way that we can have an inequality in the concentrations of H^+ and

OH^- is if some B were converted to BH, or if some BH were converted to B. This, then, gives us the stoichiometric equation

$$[H^+] = [OH^-] + [B] - C_B^\circ \tag{7-7}$$

which, on rearrangement, gives

$$[B] = C_B^\circ + [H^+] - [OH^-] \tag{7-8}$$

Noting that $c_B^\circ + c_{BH}^\circ$ must always equal $[B] + [BH]$, then

$$[BH] = c_{BH}^\circ - [H^+] + [OH^-] \tag{7-9}$$

The substitution of $K_w/[H^+]$ for $[OH^-]$ in Eqs. (7-8) and (7-9), and then substitution into Eq. (7-4) results in a cubic equation for $[H^+]$ in terms of c_B° and c_{BH}°.

Under almost all practical conditions, however, Eqs. (7-8) and (7-9) can be simplified since either $[OH^-]$ will be much greater than $[H^+]$, or $[H^+]$ will be much greater than $[OH^-]$, or both will be much smaller than both c_B° and c_{BH}°. For example, in kinetic studies where we wish to establish pseudo-first-order conditions, even if the reactant concentration were extremely low, we would certainly keep both c_B° and c_{BH}° above 10^{-3} M. Then, in the pH range from 5 to 9, $[B]$ and $[BH]$ are equal, to within 1%, to c_B° and c_{BH}°, respectively. At pH below 5, the concentration of OH^- is less than 0.01% of $[H^+]$, and at pH above 9, $[H^+]$ is less than 0.01% of $[OH^-]$. These simplifications lead to the three limiting forms of Eq. (7-4) (c_B° and $c_{BH}^\circ > 10^{-3}$ M):

$5 < pH < 9$

$$[H^+] = K_a \, c_{BH}^\circ / c_B^\circ \tag{7-10}$$

$pH < 5$

$$[H^+] = (1/2) \left\{ -(c_B^\circ + K_a) + [(c_B^\circ + K_a)^2 + 4K_a c_{BH}^\circ]^{1/2} \right\} \tag{7-11}$$

$pH > 9$

$$[H^+] = (1/2c_B^\circ) \left\{ (K_a c_{BH}^\circ + K_w) + [(K_a c_{BH}^\circ + K_w)^2 + 4c_B^\circ K_a K_w]^{1/2} \right\} \tag{7-12}$$

The use of buffers to attain pseudo-first-order conditions for acid- or base-catalyzed reactions, in addition to requiring that both components of the buffer be present at concentrations much higher than the reactant, also requires that the proton transfer from buffer to solvent, or vice versa, be much faster than the reaction under study.

7-2. SPECIFIC ACID CATALYSIS

The term "specific acid catalysis" is applied to those reactions in which the proton is the only acid whose concentration appears in the rate expression. This type of catalysis is usually characteristic of reactions in which an initial reactant is rapidly and reversibly converted to its conjugate acid, which then undergoes slow conversion by one or more steps to product.

The general mechanism for specific acid catalysis may be written as

$$S + H^+ \ \xrightarrow{\text{fast}}\ SH^+$$

$$SH^+ \ \xrightarrow[k_2]{\text{slow}}\ \text{products}$$

We define:

$$K_{SH}^\circ = \left(\frac{[S]\,[H^+]}{[SH^+]}\right)\left(\frac{\gamma_S \gamma_H}{\gamma_{SH}}\right) \tag{7-13}$$

The rate expression for the above scheme is

$$\text{Rate} = k_2^\circ\,[SH^+]\,\frac{\gamma_{SH}}{\gamma_*} \tag{7-14}$$

where γ_* is the activity coefficient of the transition state for the slow step. Before solving Eq. (7-13) for $[SH^+]$ and substituting this result into Eq. (7-14), we must pay some attention to the experimental measurement. If we wish to measure the rate of formation of product, or of disappearance of total reactant (i.e., $[S] + [SH^+]$), the direct solution of Eq. (7-13) for $[SH^+]$ in terms of $[S]$ will not give an integrable form for the rate expression in the general case.

We may, however, define the quantity C,

$$C = [S] + [SH^+] \tag{7-15}$$

and substitute for $[S]$ in Eq. (7-13) to obtain

$$[SH^+]\gamma_{SH} = \frac{C\,[H^+]\,\gamma_S \gamma_H}{(K_{SH}^\circ + [H^+]\gamma_S \gamma_H / \gamma_{SH})} \tag{7-16}$$

which reduces to

$$[SH^+]\gamma_{SH} = C\,[H^+]\,\frac{\gamma_S \gamma_H}{K_{SH}^\circ} \tag{7-17}$$

if

$$K_{SH}^{\circ} >> [H^+] \frac{\gamma_S \gamma_H}{\gamma_{SH}}$$

that is, if the concentration of SH^+ is very small in comparison to [S]. We shall assume that this condition is met, and thus arrive at the rate expression:

$$\text{Rate} = C \frac{k_2^{\circ}}{K_{SH}^{\circ}} \frac{\gamma_S a_H}{\gamma_*} \qquad (7\text{-}18)$$

where a_H is the activity of the proton. Under buffered pseudo-first-order conditions, the pseudo-first-order rate constant is

$$k_{\Psi} = k_2^{\circ} (H^+) \frac{\gamma_S \gamma_H}{K_{SH}^{\circ} \gamma_*} \qquad (7\text{-}19)$$

If the limiting conditions leading to Eq. (7-17) were not satisfied, the rate expression would be somewhat more complex, but we would still arrive at the conclusion that the pseudo-first-order rate constant is a function of only $[H^+]$. In the buffered solutions, then, as long as the buffer ratio is kept constant, the buffer component concentrations will have no effect on k_{Ψ}.

We may further note that if γ_S/γ_* were equal to the ratio of activity coefficients for some series of acids used to establish an acidity function, then log k_{Ψ} would be a linear function of H, with a slope of unity.

7-3. SPECIFIC BASE CATALYSIS

The term "specific base catalysis" is applied to those reactions in which hydroxide ion is the only base whose concentration appears in the rate expression. This type of catalysis is generally characteristic of reactions in which an initial reactant is rapidly and reversibly converted to its conjugate base, which then undergoes slow reaction to form product.

The kinetic expression for the general mechanism

$$SH + OH^- \xrightleftharpoons{\text{fast}} S^- + H_2O$$

$$S^- \xrightarrow[k_2]{\text{slow}} \text{products}$$

is derived in a manner completely analogous to that for specific acid catalysis. Defining

$$K_b^\circ = \frac{[S^-]}{[SH]\,[OH^-]}\,\frac{\gamma_S}{\gamma_{SH}\gamma_{OH}} \tag{7-20}$$

and subject to the condition that $[S^-] \ll [SH]$, we obtain

$$\text{Rate} = k_2^\circ K_b^\circ \frac{a_{OH}\gamma_{SH}}{\gamma_*}\,C \tag{7-21}$$

and

$$k_\Psi = k_2^\circ K_b^\circ \frac{a_{OH}\gamma_{SH}}{\gamma_*} \tag{7-22}$$

Again, the important point is that k_Ψ depends only on the buffer ratio in a buffered solution and not upon the concentration of the buffer. Note again, also, that if the ratio γ_{SH}/γ_* were equal to that for some series of acids used to establish an H_- acidity function, then $\log k_\Psi$ would be a linear function of H_- with a slope of unity.

In the buffer solutions used for the study of either specific acid or specific base catalysis, there will be several Brønsted acids and bases present in the solution. The proton, the buffer acid, and water are the Brønsted acids, and hydroxide ion, the buffer base, and water are Brønsted bases. Any of these acids can react with a substrate S to form its conjugate acid, and any of the bases can react with a substrate SH to form its conjugate base. As long as the reactions are fast enough to maintain equilibrium of S with SH^+, or SH with S^-, however, the ratio $[SH^+]/[S]$, or $[SH]/[S^-]$, will depend only on the concentration of the proton.

If the acid-base reactions are not fast enough to maintain equilibrium between the initial reactant and its conjugate acid, or its conjugate base, then we must consider the reaction of each Brønsted acid, or Brønsted base, present in the solution in setting up the kinetic expression for reactions.

7-4. GENERAL ACID CATALYSIS

The term "general acid catalysis" is applied to those reactions for which the kinetic expression contains the concentrations of all Brønsted acids present in the solution. Such reactions are usually characterized by having a slow proton transfer step either directly forming product, or forming the conjugate acid of the initial reactant, which then undergoes fast conversion to product.

The mechanistic scheme (where BH is the buffer acid) for general acid catalysis is

$$S + H^+ \xrightarrow{k_H} SH^+$$

$$S + H_2O \xrightarrow{k_W} SH^+ + OH^-$$

$$S + BH \xrightarrow{k_{BH}} SH^+ + B$$

$$SH^+ \xrightarrow{fast} products$$

The kinetic expression for this scheme is

$$Rate = (k_H^\circ [H^+] \ \gamma_H + k_W^\circ \gamma_W + k_{BH}^\circ [BH] \ \gamma_{BH}) \frac{[S] \gamma_S}{\gamma_*} \qquad (7\text{-}23)$$

and the pseudo-first-order rate constant is defined by

$$k_\Psi = (k_H^\circ [H^+] \ \gamma_H + k_W^\circ \gamma_W + k_{BH}^\circ [BH] \ \gamma_{BH}) \frac{\gamma_S}{\gamma_*} \qquad (7\text{-}24)$$

7-5. GENERAL BASE CATALYSIS

The mechanistic scheme (where B is the buffer base) for general base catalysis is completely analogous to that for general acid catalysis:

$$SH + OH^- \xrightarrow{k_{OH}} S^- + H_2O$$

$$SH + H_2O \xrightarrow{k_W} S^- + H_3O^+$$

$$SH + B \xrightarrow{k_B} S^- + BH$$

$$S^- \xrightarrow{fast} products$$

The pseudo-first-order rate constant is

$$k_\Psi = (k_{OH}^\circ [OH^-] \ \gamma_{OH} + k_W^\circ \gamma_W + k_B^\circ [B] \ \gamma_B) \frac{\gamma_{SH}}{\gamma_*} \qquad (7\text{-}25)$$

The activity coefficient γ_* in Eqs. (7-23)-(7-25) is actually a weighted average of the activity coefficients of the transition states for the three individual steps.

The individual rate constants in Eqs. (7-24) and (7-25) can, in principle, be obtained from buffer system studies. For example, in Eq. (7-24), k_{BH} can be obtained by varying the concentration of BH while keeping the buffer ratio constant. A plot of k_ψ vs [BH] would then have a slope of k_{BH}. Similarly, k_H can be obtained by varying the buffer ratio while keeping [BH] constant, and constructing a plot of $[H^+]$ vs k_ψ, the slope of which gives k_H. Knowing k_H and k_{BH}, k_w can then be evaluated. Completely analogous experiments allow the evaluation of the rate constants in Eq. (7-25).

7-6. ODD CASES OF ACID-BASE CATALYSIS

In addition to the two simple mechanisms above which lead to kinetic expressions containing the concentrations of all general acids, or all general bases, there are two fairly frequently encountered mechanisms which are slightly more complex. Consider the following mechanism, where HS' is an isomer of the original reactant HS:

$$SH + H^+ \xrightleftharpoons{\text{fast}} HSH^+$$

$$HSH^+ + OH^- \xrightarrow{k_{OH}} HS' + H_2O$$

$$HSH^+ + H_2O \xrightarrow{k_w} HS' + H_3O^+$$

$$HSH^+ + B \xrightarrow{k_B} HS' + BH$$

$$HS' \xrightarrow{\text{fast}} \text{products}$$

For example, HS might be a ketone and HS' the corresponding enol. Using definitions and limiting assumptions analogous to those already discussed:

$$K^\circ_{HSH} = \frac{([SH]\ [H^+]/[HSH^+])\ \gamma_{SH}\gamma_H}{\gamma_{HSH}} \tag{7-26}$$

if $[HSH^+] \ll [SH]$. Then:

$$\text{Rate} = \frac{(k^\circ_{OH}a_{OH} + k^\circ_w\gamma_w + k^\circ_B a_B)\ ([SH]/K^\circ_{HSH})\ a_H\gamma_{SH}}{\gamma_*} \tag{7-27}$$

But, since $K^\circ_w = a_H a_{OH}/\gamma_w$ and $K^\circ_a = a_B a_H/a_{BH}$, this expression can also be written

$$\text{Rate} = \frac{(k'_H[H^+]\ \gamma_H + k'_w\gamma_w + k'_{BH}[BH]\ \gamma_{BH})\ [SH]\ \gamma_{SH}}{\gamma_*} \tag{7-28}$$

where

$$k'_H = \frac{\gamma_w k^\circ_w}{K^\circ_{HSH}} \; ; \; k'_w = \frac{k^\circ_{OH} K^\circ_w}{K^\circ_{HSH}} \; ; \; k'_{BH} = \frac{k^\circ_B K^\circ_a}{K^\circ_{HSH}} \; .$$

Equation (7-28) is kinetically indistinguishable from Eq. (7-23). Thus, a mechanism involving specific acid followed by general base catalysis gives the same kinetic behavior as a mechanism involving simple general acid catalysis.

An analogous mechanism involving specific base catalysis followed by general acid catalysis will give the same kinetic behavior as a simple mechanism of general base catalysis. The proof of this statement is left as a problem at the end of this section.

Although, as we pointed out above, it is possible in principle to evaluate the individual rate constants in Eqs. (7-24), (7-25), and (7-28), and thereby establish the operation of general catalysis, difficulties are frequently encountered in practice. These difficulties arise in cases where one of the individual rate constants is so large relative to the others that it dominates the kinetics. For example, if k°_w were of numerical magnitude equal to k°_H and k°_{BH} in Eq. (7-24), then variations in [H$^+$] or in [BH] up to 10^{-2} M would produce less than a 1% variation in k_ψ, and unless extreme precision of rate measurements were attained, we would completely miss the existence of acid catalysis.

7-7. THE BRØNSTED CATALYSIS LAW [1,2]

It is frequently found that the relative rate constants for a series of acids acting as general acid catalysts in a particular reaction, or for a series of bases acting as general base catalysts in a particular reaction, are related to the equilibrium acidities of the acids, or to the equilibrium basicities of the bases, by relationships similar to the Hammett and Taft relationships. The empirical equations

$$\log k_{HB_i} = \alpha \log K_{a[HB_i]} + \text{const} \tag{7-29}$$

$$\log k_{B_i} = \beta \, pK_{a[HB_i]} + \text{const} \tag{7-30}$$

are both called the Brønsted equation or the Brønsted Catalysis Law. Equation (7-29) applies to acid catalysis, where k_{HB_i} is the catalytic rate constant for the particular acid HB$_i$, and $K_{a[HB_i]}$ is its corresponding acid ionization constant. The constant α is called the Brønsted slope for obvious reasons. Equation (7-30) applies to base catalysis and the quantities appearing are defined in a manner analogous to that for Eq. (7-29).

A fairly typical example of the applicability of Eq. (7-29) is furnished by the data for acid-catalyzed dehydration of acetaldehyde hydrate, $CH_3CH(OH)_2$, in water at 25 °C. The rate and acidity constants are shown in Table 7-1, and the plot according to Eq. (7-29) is shown in Figure 7-1.

TABLE 7-1

General Acid-Catalyzed Dehydration of $CH_3CH(OH)_2$ at 25 °C [2]

Acid	$\log k \ (M^{-1} \ sec^{-1})$	$\log K_a$
Phenol	-2.87	-9.97
p–Chlorophenol	-2.99	-9.18
o–Chlorophenol	-2.73	-8.49
m–Nitrophenol	-2.57	-8.28
2,4-Dichlorophenol	-2.43	-7.74
o–Nitrophenol	-2.25	-7.17
p–Nitrophenol	-2.06	-7.17
2,4,6-Trichlorophenol	-1.59	-6.41
Propionic acid	-0.52	-4.87
Acetic acid	-0.49	-4.75
β-Phenylpropionic acid	-0.42	-4.66
Trimethylacetic acid	-0.41	-5.03
Phenylacetic acid	-0.26	-4.31
Formic acid	-0.14	-3.75
β-Chloropropionic acid	-0.14	-3.98
Diphenylacetic acid	-0.13	-3.94
Bromoacetic acid	0.33	-2.86
Chloroacetic acid	0.39	-2.82
Cyanoacetic acid	0.56	-2.45
Dichloroacetic acid	1.10	-1.30

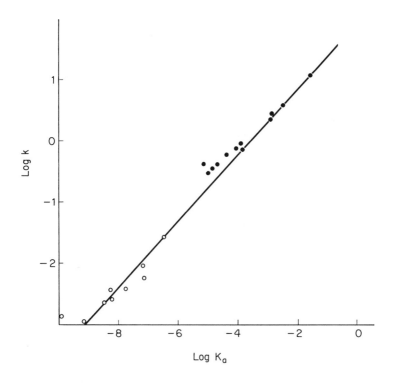

Figure 7-1. Brønsted plot of data in Table 1.

Notice particularly that the correlation is reasonably good even for those acids, such as formic acid and the o-substituted phenols, where the Hammett or Taft equations usually show substantial deviations. In the particular example here, both carboxylic acids and phenols fit a single line within reasonable limits. Frequently, however, it is found that different types of acids, or different types of bases, define different Brønsted lines when plotted according to Eqs. (7-29) or (7-30).

There are some fundamental difficulties in applying the Brønsted equation to k_W and k_H or k_{OH} due to the difficulty of defining K_a's for water and the proton [3]. In addition, we must recognize that k_W is a first-order constant, while the other rate constants are second-order. The naive resolution of this problem is to change our standard state definition for water [4,5]. Instead of defining $a_W = 1$ for pure water, we define the activity to be equal to the molarity of water in pure water, that is, $a_W = 55.5$ M. We then convert k_W to a second-order rate constant by dividing by 55.5.

A great deal of care is necessary in reading the literature, in fact, to determine whether the water rate constant is reported as a first-order (units of reciprocal time) or a second-order (units of reciprocal concentration times reciprocal time) constant. With the new definition of standard state for water, the pK_a for H_3O^+ is usually assigned the value of -1.74, and the pK_a for H_2O is assigned the value of 15.74. In actual fact, the points for H_3O^+, OH^-, and H_2O frequently fall badly enough off Brønsted plots for any consistently defined values of the above pK's that further concern is not warranted in these assignments, which nevertheless are frequently sufficient for order-of-magnitude estimates.

Consideration of Eqs. (7-29) and (7-30), using the above values of the pK's of hydronium ion, water, and hydroxide ion, shows that the most difficulty observed cases of general acid or of general base catalysis will be those which have Brønsted slopes close to either 0 or 1. If the Brønsted slope is 0, the second-order rate constant for water will be approximately the same as that for other catalysts. Since water is present at 55.5 M, however, the water-catalyzed reaction will dominate the kinetic expression as we discussed above. If the Brønsted slope is close to 1, then for acid catalysis the hydronium ion rate constant, or for base catalysis the hydroxide ion rate constant, will be much larger than those for other catalysts and will dominate the kinetic expressions. Thus, very small Brønsted slopes will make it difficult to detect any catalysis, and very large Brønsted slopes will make it difficult to distinguish between specific and general catalysis. It should come as no surprise, then, to find that most observed cases of general catalysis give Brønsted slopes of ca. 0.3 to 0.7.

The Brønsted slopes, α or β, for catalysis of a given reaction by a series of similar acids or bases are frequently interpreted as measures of the extent of proton transfer at the transition states for the reactions [6]. The reasoning behind this interpretation is fairly simple. In acid catalysis, for example, if the proton is essentially completely transferred from the acid to the reactant substrate at the transition state, then the acid moiety "looks like" the conjugate base of the acid, and any structural effects that influence the acidity of the acid should similarly influence its catalytic ability. In this case, we would expect the Brønsted slope to equal 1. If, on the other hand, the proton is still firmly attached to the acid at the transition state, then acidity should have no influence on catalytic ability and we expect a Brønsted slope of 0.

The Brønsted slopes, according to the above reasoning, should always be between 0 and 1. The reasoning implicitly assumes, however, that as the proton is transferred from the acid, or to a base, all other features of the structures of the acid or base have a monotonic dependence on the extent of transfer. For most simple oxygen or nitrogen acids and bases, this assumption appears quite reasonable. There are specific cases, however, particularly in those where resonance effects are quite different in the acid and conjugate base, where the assumption must be incorrect. For example, in proton transfers from nitro compounds,

the predominant resonance form of the conjugate base places the negative charge on the oxygens of the nitro group. It is quite possible, then, that the charge at the transition state is more strongly "felt" by the R groups than it is in the conjugate base simply because it is two atoms closer to the R groups. In fact, the Brønsted slope α for such cases has been observed to be greater than 1 [8-11].

There is no doubt that Eqs. (7-29) and (7-30) must fail if a wide enough range of acids or bases are included in the study of any reaction. Suppose we were to study a given acid-catalyzed reaction employing a series of acids whose pK's cover a wide range. Suppose further that we found a Brønsted slope of 0.5 for acids with pK's in the range from 8 to 10, and that the k_{HB} for the acid with pK = 10 were 10^7 M^{-1} sec^{-1}. Then, for an acid with pK = 2, Eq. (7-29) would predict a k_{HB} of 10^{11} M^{-1} sec^{-1}. This rate constant, however, is greater than the rate constant for the encounter of two molecules in solution. From a consideration of diffusion of molecules in solution, it is fairly simple to show that the rate at which two molecules can encounter each other will generally be governed by a diffusion rate constant of magnitude 10^9-10^{10} M^{-1} sec^{-1} for most ions or molecules [12]. In the study of a series of acids, once the diffusion-controlled rate is reached, further increases in the acidity of the acids cannot increase the rate of reaction, and, therefore, the Brønsted slope must become equal to 0.

7-8. DIFFUSION-CONTROLLED PROTON TRANSFER REACTIONS [13,14]

To simplify the situation, let us consider a simple proton transfer reaction involving a series of acids HA_i reacting with a given base B:

$$HA_i + B \rightleftharpoons BH^+ + A_i^-$$

$$K_i = \frac{K_a[HA_i]}{K_a[BH^+]} \tag{7-31}$$

The equilibrium constant for the overall reaction, K_i, is given by Eq. (7-31). The mechanism for any such reaction may be written

$$HA_i + B \xrightleftharpoons[k_{-1}]{k_1} [B \cdots HA_i] \qquad (7\text{-}32)$$

$$[B \cdots HA_i] \xrightleftharpoons[k_{-2}]{k_2} [BH^+ \cdots A_i^-] \qquad (7\text{-}33)$$

$$[BH^+ \cdots A_i^-] \xrightleftharpoons[k_{-3}]{k_3} BH^+ + A_i^- \qquad (7\text{-}34)$$

in which the species $[B \cdots HA_i]$ and $[BH^+ \cdots A_i^-]$ are simple "encounter complexes." That is, k_1 is a diffusion rate constant for the encounter of B and HA_i, and k_{-3} is a diffusion rate constant for the encounter of BH^+ and A_i^-. We assume that the diffusion coefficients for all HA_i are equal, and similarly for all A_i^-. The second step of the reaction, Eq. (7-33), is assumed to be the only step sensitive to variations in the acidities of the acids HA_i.

In the absence of specific bonding forces in the encounter complexes, the equilibrium constants for the first step, Eq. (7-32), will be on the order of 10^{-1}-10^{-2} M^{-1}. Thus, since k_1 will be approximately equal to 10^{10} M^{-1} sec^{-1}, k_{-1} is approximately 10^{11} sec^{-1}. Neglecting the coulombic interactions, we estimate $k_{-3} = k_1$, and $k_3 = k_{-1}$.

We can now apply the Bodenstein approximation to the concentrations of both encounter complexes, and arrive at the kinetic expression

$$\frac{d[BH^+]}{dt} = \frac{k_1 k_2 k_3 [HA_i][B] - k_{-1} k_{-2} k_{-3} [BH^+][A_i^-]}{k_2 k_3 + k_{-1} k_{-2} + k_{-1} k_3} \qquad (7\text{-}35)$$

from which it is easy to see that the forward and reverse rate constants observed are

$$k_f = \frac{k_1 k_2 k_3}{k_2 k_3 + k_{-1} k_{-2} + k_{-1} k_3} \qquad (7\text{-}36)$$

$$k_r = \frac{k_{-1} k_{-2} k_{-3}}{k_2 k_3 + k_{-1} k_{-2} + k_{-1} k_3} \qquad (7\text{-}37)$$

Three limiting cases can arise, each corresponding to a particular one of the three steps being rate limiting. If HA_i is a very weak acid such that $K_i \ll 1$, we may expect that $k_2 \ll k_{-1}$ and $k_{-2} \gg k_3$. Therefore, Eqs. (7-36) and (7-37) become

$$k_f = K_1 K_2 k_3$$

$$k_r = k_{-3} \qquad (7\text{-}38)$$

For moderately strong acids HA_i, we expect $k_2 \ll k_{-1}$ and $k_{-2} \ll k_3$, which gives

$$k_f = K_1 k_2$$
$$k_r = k_{-2} K_3 \qquad\qquad (7\text{-}39)$$

Finally, for very strong acids for which $K_i \gg 1$, we expect $k_2 \gg k_{-1}$ and $k_3 \gg k_{-2}$, giving

$$k_f = k_1$$
$$k_r = k_{-1} K_2 K_3 \qquad\qquad (7\text{-}40)$$

Equations (7-38), (7-39), and (7-40) are of course the expressions corresponding to the cases where step 3, step 2, and step 1, respectively, are rate determining.

Suppose now that k_2 obeys the Brønsted equation (7-29) with $\alpha = 0.5$, and that k_{-2} obeys Eq. (7-30) with $\beta = 0.5$. If we were to plot log k_f vs log K_i, we would find three portions in the curve. For large negative values of log K_i, Eq. (7-38) would apply, and the slope of the plot would be 1. For moderate values of K_i, Eq. (7-39) would apply, and we would have a slope of 0.5. For large positive values of log K_i, Eq. (7-40) would apply, and the slope would be 0. If log k_r were plotted vs log K_i, in the range where the plot of log k_f has a slope of 1, the plot of log k_r must have a slope of 0; in the range where the log k_f plot has a slope of 0, the log k_r plot must have slope of -1; and in the range where the log k_f plot has slope of α, the log k_r plot must have slope of $1 - \alpha$. The situation is sketched in Figure 7-2.

The entire question of the range of applicability of Brønsted plots, then, depends on the magnitudes of k_2 or k_{-2} relative to the rate constants for dissociation of the encounter complexes. These magnitudes are determined by the activation barriers for step 2, Eq. (7-33), in the forward and reverse directions.

We will return to this question of barriers in the actual proton transfer step, and some of the interesting chemistry which results when the encounter complex $[BH^+ \dots A^-]$ can rearrange faster than it can dissociate, in a later chapter. It is sufficient here to point out that most simple oxygen or nitrogen acids and bases are found to have extremely small barriers to the proton transfer step, and that step 2 is seldom rate determining in these cases. The changeover from slope of unity to slope of zero in a plot such as Figure 7-2 for these acids and bases occurs over a very narrow range of log K_i, and either k_f or k_r is almost always quite close to the diffusion-controlled limit [6,7,13,14].

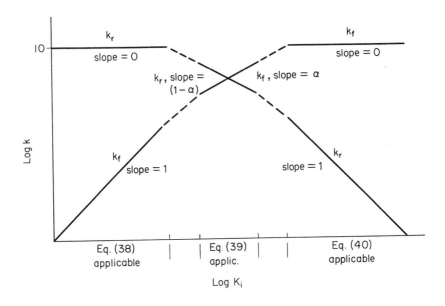

Figure 7-2. The Eigen modification of the Brønsted relationship.

7-9. BELL-SHAPED pH-RATE PROFILES [15]

In all of the cases of acid or base catalysis that we have discussed so far, we have seen that the pseudo-first-order rate constants are linear functions of $[H^+]$. There are many cases, however, in which the pseudo-first-order rate constants for reaction exhibit bell-shaped curves when plotted against $[H^+]$; that is, k_ψ increases with increasing $[H^+]$ at high pH, and decreases with increasing $[H^+]$ at low pH, or vice versa.

One thoroughly studied case of such behavior is the reaction of a ketone with hydroxylamine to form the oxime [15]. A possible mechanism of this reaction is the following:

$$\begin{array}{c}
\underset{\substack{R' \\ \text{C}}}{\overset{R}{\diagdown}}C=O + NH_2OH \underset{k_{-1}}{\overset{k_1}{\rightleftharpoons}} \underset{I}{\overset{OH}{\underset{NHOH}{R-\overset{|}{\underset{|}{C}}-R}}}
\end{array}$$

$$\underset{I}{\overset{OH}{\underset{NHOH}{R-\overset{|}{\underset{|}{C}}-R}}} \underset{k_{-2}}{\overset{k_2\,[H^+]}{\rightleftharpoons}} \underset{IH^+}{\overset{OH_2^+}{\underset{NHOH}{R-\overset{|}{\underset{|}{C}}-R}}} \overset{k_3}{\longrightarrow} \underset{Ox}{oxime + H_3O^+}$$

Application of the Bodenstein approximation to the concentrations of both intermediates, [I] and [IH$^+$], leads to the kinetic expression:

$$\text{Rate} = \frac{k_1 k_2 k_3}{k_2 k_3 [H^+] + k_{-1}(k_{-2} + k_3)} \, [C] \, [N] \, [H^+] \tag{7-41}$$

Before defining a pseudo-first-order rate constant, however, we must take into account the fact that hydroxylamine undergoes an acid-base reaction:

$$HONH_3^+ \underset{}{\overset{K_a}{\rightleftharpoons}} HONH_2 + H^+$$

Defining $N_0 = (HONH_3^+) + (HONH_2)$, we find

$$[N] = \frac{N_0}{1 + [H^+]/K_a} \tag{7-42}$$

Substitution of Eq. (7-42) into (7-41) then gives

$$\text{Rate} = \frac{k_1 k_2 k_3 K_a N_0}{[k_2 k_3 (H^+) + k_{-1}(k_{-2} + k_3)] \, [1 + K_a/(H^+)]} \, [C]$$

$$= k_\Psi [C] \tag{7-43}$$

where the last equality defines the pseudo-first-order rate constant, k_Ψ.

At very high pH, $K_a \gg [H^+]$, and $k_{-1}(k_{-2} + k_3) \gg k_2 k_3 [H^+]$, and Eq. (7-43) gives

$$k_\Psi = \frac{k_1 k_2 k_3 N_0}{k_{-1}(k_{-2} + k_3)} \, [H^+] \tag{7-44}$$

Under these conditions, either step 2 or step 3 of the reaction is rate determining, and k_Ψ is directly proportional to $[H^+]$.

As the pH is lowered, we will eventually reach a point where $K_a \ll [H^+]$ and $k_{-1}(k_{-2} + k_3) \ll k_2 k_3 [H^+]$. Under these conditions, Eq. (7-43) becomes

$$k_\Psi = \frac{k_1 K_a N_0}{[H^+]} \tag{7-45}$$

which shows that step 1 has become rate determining, and that k_Ψ is inversely proportional to $[H^+]$.

A complete plot of Eq. (7-43) for a hypothetical case where $K_a = 1.0 \times 10^{-6}$, $k_1 = 2.0 \times 10^5$ M^{-1} sec^{-1}, $k_{-1}(k_{-2} + k_3)/k_2 k_3 = 1.0 \times 10^{-7}$ is shown in Figure 7-3. In the actual case of the reactions of ketones with hydroxylamine, the experimental points at low pH are appreciably above the

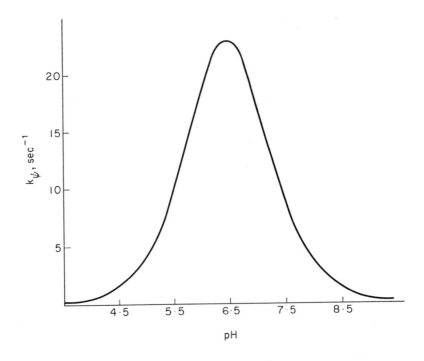

Figure 7-3. Plot of Eq. (7–43), $N_0 = 2 \times 10^{-4}$ M.

symmetrical curve predicted by Eq. (7–43) and clearly indicate an acid-catalyzed contribution to the formation of the intermediates, I or IH^+, in the above mechanism.

The reactions of carbonyl compounds furnish a rich variety of acid- and base-catalyzed behavior of considerable importance in the study of biochemical systems. The interested reader is referred to Jencks' work [4,5,15].

The entire discussion in this chapter has dealt with kinetic studies of acid and base catalysis, and has emphasized the use of buffer systems for the detection of general base or general acid catalysis. The key question in these studies, of course, is whether or not a proton is being transferred in the rate-determining step of the reaction. There are many cases where the kinetic studies of the sort discussed here are either not possible, or where they do not give the desired distinction between specific and general catalysis. The study of deuterium isotope effects on rates of reaction involving proton transfers provides an alternative, and sometimes more definitive, approach to the problem.

PROBLEMS

1. Show that the following reaction scheme gives a kinetic expression which is indistinguishable from that for general base catalysis:

$$SH + OH^- \xrightleftharpoons{fast} S^- + H_2O$$

$$S^- + H_2O \xrightarrow{k_W} products$$

$$S^- + BH \xrightarrow{k_{BH}} products$$

$$S^- + H_3O^+ \xrightarrow{k_H} products$$

2. The reaction

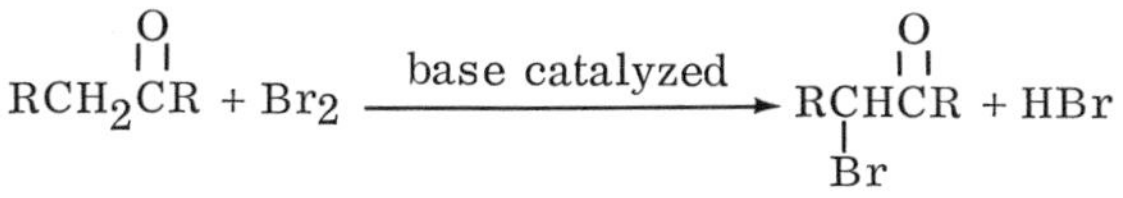

$$\underset{\substack{| \\ RCH_2CR}}{\overset{\substack{O \\ ||}}{}} + Br_2 \xrightarrow{\text{base catalyzed}} \underset{\substack{| \\ RCHCR \\ | \\ Br}}{\overset{\substack{O \\ ||}}{}} + HBr$$

was studied in buffer solutions. The following data were obtained using $H_2PO_4^-$, $HPO_4^=$ ($pK_{H_2PO_4^-} = 7.12$) buffers at an ionic strength of 0.1 M.

$[HPO_4^=]$	$[H_2PO_4^-]$	k_ψ (sec^{-1})
2.0 x 10^{-3}	4.0 x 10^{-3}	2.05 x 10^{-2}
2.0 x 10^{-3}	2.0 x 10^{-3}	2.46 x 10^{-2}
2.0 x 10^{-3}	1.0 x 10^{-3}	3.30 x 10^{-2}
2.0 x 10^{-3}	5.0 x 10^{-4}	4.96 x 10^{-2}
3.0 x 10^{-3}	3.0 x 10^{-3}	2.78 x 10^{-2}
4.0 x 10^{-3}	4.0 x 10^{-3}	3.10 x 10^{-2}
6.0 x 10^{-3}	6.0 x 10^{-3}	3.73 x 10^{-2}
1.0 x 10^{-2}	1.0 x 10^{-2}	4.99 x 10^{-2}
1.0 x 10^{-2}	2.0 x 10^{-2}	4.58 x 10^{-2}
1.0 x 10^{-2}	5.0 x 10^{-3}	5.82 x 10^{-2}

Calculate the rate constants for each catalytic base. Do the data obey the Brønsted equation?

3. Show that the following data on the rate of formation of acetone oxime are consistent with the mechanism:

$$(CH_3)_2C=O + NH_2OH \underset{k_{-1}}{\overset{k_1}{\rightleftharpoons}} (CH_3)_2C{\overset{OH}{\underset{NHOH}{\big\langle}}}$$

$$H^+ + (CH_3)_2C=O + NH_2OH \underset{k'_{-1}}{\overset{k'_1}{\rightleftharpoons}} (CH_3)_2C{\overset{OH}{\underset{NHOH}{\big\langle}}} + H^+$$

$$H^+ + (CH_3)_2C{\overset{OH}{\underset{NHOH}{\big\langle}}} \overset{k_2}{\longrightarrow} (CH_3)_2C=NOH + H_3O^+$$

Evaluate as many of the rate constants as possible. $pK_{HONH_3}^+ = 6.0$; $[NH_2OH]_0 = 0.0167$ M; $[acetone]_0 = 1.0 \times 10^{-4}$ M.

pH	$10^2 k_\psi$ (sec^{-1})	pH	$10^2 k_\psi$ (sec^{-1})
2.0	0.548	5.6	2.18
2.2	0.693	5.8	1.87
2.4	0.891	6.0	1.53
2.6	1.14	6.2	1.18
2.8	1.44	6.4	0.871
3.0	1.76	6.6	0.614
3.2	2.07	6.8	0.418
3.4	2.34	7.0	0.278
3.6	2.55	7.2	0.181
3.8	2.71	7.4	0.117
4.0	2.81	7.6	0.075
4.2	2.87	7.8	0.048
4.4	2.89	8.0	0.030
4.6	2.89	8.2	0.019
4.8	2.84	8.4	0.012
5.0	2.76	8.6	0.0077
5.2	2.63	8.8	0.0048

4. "Nucleophilic catalysis" is frequently observed in reactions of carbonyl compounds. The simple mechanism for such catalysis involves the following steps (NH is any Lewis base):

$$
\underset{\substack{\|\\ O}}{R\diagdown C\diagup X} + NH \underset{k_{-NH}}{\overset{k_{NH}}{\rightleftharpoons}} \underset{\substack{\diagup \diagdown \\ X \quad NH^+}}{R\diagdown C\diagup O^-}
$$

$$
\underset{\substack{\diagup \diagdown \\ X \quad NH^+}}{R\diagdown C\diagup O^-} \underset{k_{-2}}{\overset{k_2}{\rightleftharpoons}} \underset{\substack{\|\\ O}}{R\diagdown C\diagup N} + HX
$$

$$
\underset{\substack{\|\\ O}}{R\diagdown C\diagup N} + H_2O \xrightarrow{\text{fast}} RCOOH + NH
$$

a. Derive the kinetic expression for the reaction as written.
b. Derive the kinetic expression for a case where step 1 is specific acid catalyzed.
c. Derive the kinetic expression for the case where step 2 involves specific acid, followed by general base, catalysis.
d. Derive the kinetic expression for a case where NH may be protonated to form an inactive reagent NH_2^+ in an equilibrium.

REFERENCES

1. R. P. Bell, The Proton in Chemistry, 2nd ed., Cornell University Press, Ithaca, New York, 1973, Chap. 10.

2. A. A. Frost and R. G. Pearson, Kinetics and Mechanism, 2nd ed., John Wiley and Sons, Inc., New York, New York, 1961, pp 213-223.

3. E. M. Arnett, Prog. Phys. Org. Chem. 1, 223 (1963).

4. W. P. Jencks, Chem. Revs., 72, 705 (1972).

5. W. P. Jencks, Catalysis in Chemistry and Biochemistry, McGraw-Hill Book Co., New York, New York, 1969.

6. A. J. Kresge, Chem. Soc. Revs. 2, 475 (1973).

7. R. P. Bell, The Proton in Chemistry, 2nd ed., Cornell University Press, Ithaca, New York, 1973, pp. 214ff.

8. F. G. Bordwell, W. J. Boyle, Jr., J. A. Hautala, and K. C. Yee, J. Amer. Chem. Soc., 91, 4002 (1969).

9. F. G. Bordwell, W. J. Boyle, Jr., and K. C. Yee, J. Amer. Chem. Soc., 92, 5926 (1970).

10. F. G. Bordwell and W. J. Boyle, Jr., J. Amer. Chem. Soc., 93, 511 (1971).

11. F. G. Bordwell and W. J. Boyle, Jr., J. Amer. Chem. Soc., 94, 3907 (1972).

12. E. F. Caldin, Fast Reactions in Solution, Blackwell Scientific Publishers, Oxford, England, 1964. See, particularly pp. 10-13.

13. M. Eigen, Angew. Chem., International Ed., 3, 1 (1964).

14. C. D. Ritchie in Solvent–Solute Interactions (J. F. Coetzee and C. D. Ritchie, eds.), Marcel Dekker, Inc., New York, New York, 1969, Chap. 4.

15. W. P. Jencks, Prog. Phys. Org. Chem., 2, 63 (1964).

Chapter 8

EQUILIBRIUM AND SECONDARY KINETIC ISOTOPE EFFECTS

8-1. INTRODUCTION

In the discussion of quantum mechanics in Chapter 5, it was stated that the Born-Oppenheimer approximation allows us to calculate electronic energies of a molecule without consideration of nuclear motion. The resulting electronic Schrödinger equation contains no terms dependent on the masses of the nuclei. It follows that the electronic energy of a molecule is not affected by the substitution of one isotope for another in the molecule.

Nevertheless, the free energies, enthalpies, and entropies of a system of molecules are generally changed by isotopic substitution in the component molecules. This change in thermodynamic quantites occurs because all molecules are in motion, even at absolute zero of temperature, and the kinetic energies of these motions are mass dependent. In order to understand the effect of isotopic substitution on either equilibria or rates of chemical reactions, it is first necessary to understand how the motions of a molecule are involved in the thermodynamic properties of a system of these molecules. This is the subject matter of statistical thermodynamics.

8-2. BOLTZMANN DISTRIBUTION APPLIED TO AN ISOMERIZATION REACTION

In the following discussion, we shall start with the idea of a Boltzmann distribution and develop statistical mechanical expressions appropriate to an isomerization reaction [1]. This development will bring out the important principles necessary to the understanding of isotope effects. The treatment of other types of reactions, particularly those involving a change in the number of molecules on going from reactants to products, would require more time and space than justified for the present purposes [2].

The Boltzmann distribution law [3] [Eq. (8-1)]

$$n_i = \frac{N_0}{P} \exp(-\frac{\epsilon_i}{kT})$$ (8-1)

states that the number of particles n_i having a given energy ϵ_i is an exponential function of ϵ_i. In Eq. (8-1), k is the Boltzmann constant, T is the absolute temperature, N_0 is the total number of particles in the system, and P is a normalization constant, chosen such that $\Sigma_i n_i = N_0$. The Boltzmann distribution law is derived from a consideration of the most probable distribution of a large number of particles of total energy $E = \Sigma_i n_i \epsilon_i$. Simple statistics show that the Boltzmann distribution for such a large number of particles is overwhelmingly the most probable, and can, therefore, be considered as "the" distribution.

If the particles referred to in the above paragraph are molecules, the energies ϵ_i will be the quantum energy levels associated with various types of motions of the molecules as well as the electronic energies. Limiting our attention to a dilute gas, the energies will consist of electronic, vibrational, rotational, and translational energies of a molecule:

$$\epsilon_i = \epsilon_{j}(\text{elec}) + \epsilon_{k}(\text{vib}) + \epsilon_{l}(\text{rot}) + \epsilon_{m}(\text{trans}) \tag{8-2}$$

If the atoms in a molecule can exist in different arrangements, with the different arrangements having different quantum energy levels, the Boltzmann distribution law still applies at equilibrium. That is, the only factors which affect the rearrangement of one molecule into another at equilibrium are the energy levels for the two arrangements. In an isomerization reaction,

$$A \rightleftharpoons A'$$

the molecule (arrangement) A will have a set of energy levels ϵ_i and the molecule A' will have a different set of energy levels ϵ'_j, due to different electronic energies, vibrational frequencies, and moments of inertia of A and A'. Both of these sets of energy levels, however, are available to the system, and at equilibrium, each level of each set will be populated according to the Boltzmann distribution law. In this sense, the only distinction between A and A' is whether a particular collection of atoms finds itself in the set of levels ϵ_i or in the set ϵ'_j.

The total number of molecules having the arrangement corresponding to molecule A, n_A, is simply the sum of the Boltzmann populations of the ϵ levels:

$$n_A = \frac{N_0}{P} \sum_i \exp\left(-\frac{\epsilon_i}{kT}\right) \tag{8-3}$$

Similarly, the number of molecules having the arrangement corresponding to A', $n_{A'}$, is

$$n_{A'} = \frac{N_0}{P} \sum_j \exp\left(-\frac{\epsilon_j^!}{kT}\right) \tag{8-4}$$

The equilibrium constant for the isomerization reaction is simply the ratio of $n_{A'}$ and n_A:

$$K_i = \frac{n_{A'}}{n_A} = \frac{\sum_j \exp\left(-\frac{\epsilon_j^!}{kT}\right)}{\sum_i \exp\left(-\frac{\epsilon_i}{kT}\right)} \tag{8-5}$$

This simple result is sufficient to allow us to derive all of the thermodynamic quantities associated with the isomerization reaction in terms of the Boltzmann populations from the following known relationships:

$$\Delta G^\circ = -RT \ln K_i \tag{8-6}$$

$$\Delta H^\circ = -R \frac{d \ln k_i}{d(1/T)} \tag{8-7}$$

$$\Delta S^\circ = R \ln K_i - \frac{R}{T} \frac{d \ln k_i}{d(1/T)} \tag{8-8}$$

Considerations of other types of reactions lead to equations quite similar to Eq. (8-5) [2,3]. For the general reaction

$$A + B \rightleftharpoons C + D$$

the equilibrium constant is given by Eq. (8-9):

$$K = \frac{\overset{C}{\underset{i}{\sum}} \exp(-\epsilon_i/kT) \overset{D}{\underset{j}{\sum}} \exp(-\epsilon_j/kT)}{\overset{A}{\underset{k}{\sum}} \exp(-\epsilon_k/kT) \overset{B}{\underset{l}{\sum}} \exp(-\epsilon_l/kT)} \tag{8-9}$$

8-3. RESOLUTION OF BOLTZMANN SUMS

The summations appearing in Eqs. (8-5) and (8-9) are called "sums over states" for obvious reasons. These sums can be written in a more informative and useful manner in those cases where the energy levels for one type of motion are independent of those for another type of motion; that is, in those cases where the terms on the right-hand side of Eq. (8-2) are independent of one another. This will be the case for most molecules and temperatures of interest to the organic chemist.

Usually we will need only to consider the ground electronic state of a molecule since the excited electronic states are so high in energy that at normal temperatures the Boltzmann populations of the excited states will be negligible.

We shall see in the following discussion that the translational energy levels for a molecule depend only on the mass of the molecule and the volume of the container in which it is constrained. The translational levels, then, are independent of the amount of vibrational or rotational energy of the molecule.

It is conceivable that vibrational and rotational energy levels of a molecule would be interdependent. For example, if a molecule were rotating so rapidly that centrifugal forces became comparable in magnitude to vibrational force constants, or if a molecule were vibrating so energetically that the amplitude of vibration changed the moment of inertia of the molecule, the two types of energies could be dependent. Again, however, at reasonable temperatures for normal molecules, we expect the vibrational and rotational levels to be independent.

Under these conditions, the Boltzmann sums may be expanded by the use of Eq. (8-2) into electronic, vibrational, rotational, and translational terms:

$$\sum_i \exp\left(\frac{-\epsilon_i}{kT}\right) = \exp\left(\frac{-\epsilon_0^\circ}{kT}\right)\left\{\sum_v^{vib}\exp\left(\frac{-\epsilon_v}{kT}\right)\sum_r^{rot}\exp\left(\frac{-\epsilon_r}{kT}\right)\sum_t^{trans}\exp\left(\frac{-\epsilon_t}{kT}\right)\right\}$$

$$= \exp\left(\frac{-\epsilon_0^\circ}{kT}\right) f'_{vib}f_{rot}f_{trans} \tag{8-10}$$

where the final equality defines the f's as Boltzmann sums over the particular types of energy levels, vibrational, rotational, and translational, and ϵ_0° is the ground state electronic energy of the molecule.

8-4. THE EVALUATION OF BOLTZMANN SUMS

Each of the sums in Eq. (8-10) can, at least in principle, be evaluated from experimental measurements [4]. The vibrational energy levels can be obtained from infrared and Raman spectra; the rotational energy levels can be obtained from infrared fine structure or from microwave spectra; and the translational energy levels can be obtained from classical mechanical considerations. These energy levels can also be obtained from quantum mechanical calculations, and these give useful information as to how the energy levels depend on the structures of the molecule. This information is particularly helpful in understanding isotope effects on equilibria.

8-4.1. The Evaluation of f_{trans}

If we consider a particle of mass m moving in a box of dimensions a x b x c = V, the potential energy within the box is constant, and the Schrödinger equation becomes

$$\frac{-\hbar^2}{2m} \left(\frac{\partial^2}{\partial x^2} + \frac{\partial^2}{\partial y^2} + \frac{\partial^2}{\partial z^2}\right) \Psi_t = \epsilon_t \Psi_t \tag{8-11}$$

Since the molecule must be within the box, Ψ_t must go to 0 for $0 > x > a$, $0 > y > b$, and for $0 > z > c$, and, since Ψ_t must be continuous, it must be equal to 0 when the equality signs apply. If we assume that the wavefunction within the box can be written as a product of functions of x, y, and z, respectively, then Eq. (8-11) can be written as three independent equations [5]:

$$\frac{-\hbar^2}{2m} \frac{\partial^2 \Psi_x}{\partial x^2} = \epsilon_x \Psi_x \tag{8-12a}$$

$$\frac{-\hbar^2}{2m} \frac{\partial^2 \Psi_y}{\partial y^2} = \epsilon_y \Psi_y \tag{8-12b}$$

$$\frac{-\hbar^2}{2m} \frac{\partial^2 \Psi_z}{\partial z^2} = \epsilon_z \Psi_z \tag{8-12c}$$

$$\Psi_t = \Psi_x \Psi_y \Psi_z \qquad \epsilon_t = \epsilon_x + \epsilon_y + \epsilon_z$$

It is obvious on inspection that either of the functions $\sin \alpha x$ or $\cos \alpha x$ satisfy Eq. (8-12a). The cosine function is unsatisfactory, however, since it does not meet the requirement that the function equal 0 at x = 0. The sine function does meet this requirement, and will meet the further requirement that it be 0 at x = a, if we choose α such that

$$\alpha = \frac{n_x \pi}{a} \tag{8-13}$$

where n_x can be any integer. Therefore, the function

$$\Psi_x = A \sin \frac{n_x \pi x}{a} \tag{8-14}$$

is a general solution of Eq. (8-12a). Substitution of Eq. (8-14) into Eq. (8-12a) gives

$$\frac{\partial^2 \Psi_x}{\partial x^2} = -A(\frac{n_x \pi}{a})^2 \sin \frac{n_x \pi x}{a} = \frac{-2m}{\hbar^2} \epsilon_x A \sin \frac{n_x \pi x}{a} \tag{8-15}$$

from which we see that

$$\epsilon_x = \frac{n_x^2 h^2}{8ma^2} \tag{8-16a}$$

Completely analogous treatment of Eqs. (8-12b) and (8-12c) gives

$$\epsilon_y = \frac{n_y^2 h^2}{8mb^2} \tag{8-16b}$$

$$\epsilon_z = \frac{n_z^2 h^2}{8mc^2} \tag{8-16c}$$

Combination of the last three equations then gives the result

$$\epsilon_t = \frac{h^2}{8m} (\frac{n_x^2}{a^2} + \frac{n_y^2}{b^2} + \frac{n_z^2}{c^2}) \tag{8-17}$$

where n_x, n_y, and n_z can take on any integer values.
Substitution of Eq. (8-17) into the Boltzmann sum gives

$$f_{trans} = \sum_{n_x = 0}^{\infty} \sum_{n_y = 0}^{\infty} \sum_{n_z = 0}^{\infty} \exp \left[\frac{-h^2}{8mkT} (\frac{n_x^2}{a^2} + \frac{n_y^2}{b^2} + \frac{n_z^2}{c^2}) \right]$$

$$= \left[\sum_{n_x = 0}^{\infty} \exp (\frac{-n_x^2 h^2}{8mkTa^2}) \right] \left[\sum_{n_y = 0}^{\infty} \exp (\frac{-n_y^2 h^2}{8mkTb^2}) \right]$$

$$\left[\sum_{n_z = 0}^{\infty} \exp (\frac{-n_z^2 h^2}{8mkTc^2}) \right] \tag{8-18}$$

The translational energy levels given in Eq. (8-17) for molecules of any reasonable mass are so closely spaced that we may replace the summations over quantum numbers by integrations over the quantum numbers:

$$\sum_{n_x = 0}^{\infty} \exp(\frac{-n_x^2 h^2}{8mkTa^2}) = \int_0^{\infty} \exp(\frac{-n_x^2 h^2}{8mkTa^2}) \, dn_x \tag{8-19}$$

The integral in Eq. (8-19) can be found in any table of standard integrals, and gives the following result:

$$\sum_{n_x = 0}^{\infty} \exp(\frac{-n_x^2 h^2}{8mkTa^2}) = \frac{a(2\pi mkT)}{h}^{1/2} \tag{8-20}$$

The integrations over the quantum numbers n_y and n_z give analogous expressions, and substitution of these results into Eq. (8-18) gives

$$f_{trans} = \frac{(2\pi mkT)^{3/2}V}{h^3} \tag{8-21}$$

8-4.2. The Evaluation of f'_{vib}

Any nonlinear molecule containing n nuclei has 3n−6 vibrational modes, and each of these will produce a set of vibrational energy levels [5]. Thus, f'_{vib} will be the product of 3n−6 Boltzmann terms:

$$f'_{vib} = \sum_i \exp(\frac{-\epsilon_i}{kT}) = \prod_{j=1}^{3n-6} \left[\sum_{i=0}^{\infty} \exp(\frac{-\epsilon_{ij}}{kT}) \right] \tag{8-22}$$

where ϵ_{ij} is the energy of the i-th level of the j-th vibrational mode. For most molecules at ordinary temperatures, each of the vibrational modes can be treated as one-dimensional harmonic oscillators. That is, each mode may be treated independently using the Hamiltonian operator for the one-dimensional oscillator:

$$H_j = \frac{(-\hbar^2/2m_r)\partial^2}{\partial x_j^2} + \frac{k_j x_j^2}{2} \tag{8-23}$$

where x_j is the displacement from equilibrium, k_j is the Hooke's law force constant, and m_r is the reduced mass for the j-th vibrational mode. For a two-particle system, with masses m_a and m_b, the reduced mass is:

$$m_r = \frac{m_a m_b}{m_a + m_b} \tag{8-24}$$

In our earlier discussion of quantum mechanics, we saw that the function

$$F = (2\alpha/\pi)^{1/4} \exp(-\alpha x_j^2) \tag{8-25}$$

is an eigenfunction of H_j if

$$\alpha = \frac{(k_j m_r)^{1/2}}{2\hbar} \tag{8-26}$$

and has the eigenvalue

$$\epsilon_{jo} = \frac{\hbar}{2} \left(\frac{k_j}{m_r}\right)^{1/2} = \frac{h\nu_j}{2} \tag{8-27}$$

where we have defined ν_j according to Eq. (8-28):

$$\nu_j = \frac{(k_j/m_r)^{1/2}}{2\pi} \tag{8-28}$$

We now show that any function obtained by operating on the function F with the operator R^n

$$R = \frac{\partial}{\partial x_j} - 2\alpha x_j$$

$$R^n = \left(\frac{\partial}{\partial x_j} - 2\alpha x_j\right)^n \tag{8-29}$$

is also an eigenfunction of H_j with eigenvalue $(n + 1/2)h\nu_j$.

To prove this statement, consider operating on any function of x_j, $G(x_j)$, with the operator H_jR:

$$H_jRG(x_j) = \left[\frac{(-\hbar^2/2m_r)\partial^2}{\partial x_j^2} + \frac{k_j x_j^2}{2}\right]\left(\frac{\partial G}{\partial x_j} - 2\alpha x_j G\right)$$

$$H_jRG(x_j) = \frac{-\hbar^2}{2m_r}\left[\frac{\partial^2}{\partial x_j^2} - 4\alpha^2 x_j^2\right]\left[\frac{\partial G}{\partial x_j} - 2\alpha x_j G\right] \tag{8-30}$$

$$= \frac{-\hbar^2}{2m_r}\left[\frac{\partial^3 G}{\partial x_j^3} - 2\alpha\frac{\partial(G + x_j\frac{\partial G}{\partial x_j})}{\partial x_j} - 4\alpha^2 x_j^2\frac{\partial G}{\partial x_j} + 8\alpha^3 x_j^3 G\right]$$

$$= \frac{-\hbar^2}{2m_r}\left[\frac{\partial^3 G}{\partial x_j^3} - 4\alpha\frac{\partial G}{\partial x_j} - 2\alpha x_j\frac{\partial^2 G}{\partial x_j^2} - 4\alpha^2 x_j^2\frac{\partial G}{\partial x_j} + 8\alpha^3 x_j^3 G\right]$$

Now consider operating on the same general function $G(x_j)$ with the operator RH_j:

$$RH_jG = \left[\frac{\partial}{\partial x_j} - 2\alpha x_j\right]\left[\frac{-\hbar^2}{2m_r}\frac{\partial^2 G}{\partial x_j^2} + k_j x_j^2\frac{G}{2}\right] \tag{8-31}$$

$$= \frac{-\hbar^2}{2m_r} \left[\frac{\partial^3 G}{\partial x_j^3} - 8\alpha^2 x_j G - 4\alpha^2 x_j^2 \frac{\partial G}{\partial x_j} - 2\alpha x_j \frac{\partial^2 G}{\partial x_j^2} + 8\alpha^3 x_j^3 G \right]$$

Subtraction of Eq. (8-31) from Eq. (8-30) gives us the following equation:

$$[H_j R - RH_j] G(x_j) = \frac{2\alpha\hbar^2}{m_r} \left(\frac{\partial}{\partial x_j} - 2\alpha x_j \right) G(x_j)$$

$$= \frac{2\alpha\hbar^2}{m_r} RG(x_j) \qquad\qquad (8-32)$$

Since Eq. (8-32) has been derived for any function of x_j, we may write it as an operator equation:

$$H_j R - RH_j = (2\alpha h^2/m_r) R \qquad\qquad (8-33)$$

Substitution from Eqs. (8-26) and (8-28) gives

$$H_j R - RH_j = h\nu_j R$$

or

$$RH_j = H_j R - h\nu_j R \qquad\qquad (8-34)$$

From our previous knowledge that F is an eigenfunction of H_j, and using Eq. (8-27), we may write

$$H_j F = [h\nu_j/2] F \qquad\qquad (8-35)$$

Left-multiplication of Eq. (8-35) by the operator R gives

$$RH_j F = (h\nu_j/2) RF \qquad\qquad (8-36)$$

and substitution from Eq. (8-34) then gives

$$(H_j R - h\nu_j R) F = (h\nu_j/2) RF$$

or

$$H_j (RF) = h\nu_j (1 + 1/2) RF \qquad\qquad (8-38)$$

which shows that the function RF is an eigenfunction of H_j, with eigenvalue of $(1 + 1/2)h\nu_j$. It should now be clear that if we left-multiply Eq. (8-38) by R, and substitute again from Eq. (8-34), that we will find that $R^2 F$ is an eigenfunction of H_j with eigenvalue $(2 + 1/2)h\nu_j$.

It is then obvious that $R^n F$ is an eigenfunction of H_j with eigenvalue $(n + 1/2)h\nu_j$. Thus, the energy levels for the harmonic oscillator are

$$\epsilon_n = (n + 1/2)h\nu_j \tag{8-39}$$

where n is any integer. Although we will not prove the statement here, it is also true that these are the only eigenvalues of H_j. (See problem 3 at the end of this chapter.)

With the result of Eq. (8-39), we may now return to Eq. (8-22). Substitution gives

$$f'_{vib} = \prod_{j=1}^{3n-6} \sum_{n=0}^{\infty} \exp\left[\frac{-(n_j + 1/2)h\nu_j}{kT}\right] \tag{8-40}$$

Since $h\nu_j/2$ of energy will always be present for each vibration, we may factor out these terms and define a new quantity, f_{vib}:

$$f_{vib} = \prod_{j=1}^{3n-6} \sum_{n=0}^{\infty} \exp(\frac{-n_j h\nu_j}{kT}) \tag{8-41}$$

and a quantity ϵ_{zp}:

$$\epsilon_{zp} = \sum_{j=1}^{3n-6} \frac{h\nu_j}{2} \tag{8-42}$$

We may further note that each of the sums in Eq. (8-41) can be written in an alternate form:

$$\sum_{n=0}^{\infty} \exp[\frac{-n_j h\nu_j}{kT}] = [1 - \exp(\frac{-h\nu_j}{kT})]^{-1} \tag{8-43}$$

Equation (8-43) can be most easily verified by multiplying both sides by the quantity $[1 - \exp(-h\nu_j/kT)]$.

Substitution of Eq. (8-43) into (8-41) gives

$$f_{vib} = \prod_{j=1}^{3n-6} [1 - \exp(\frac{-h\nu_j}{kT})]^{-1} \tag{8-44}$$

and Eq. (8-40) may then be written

$$f'_{vib} = \exp\left(\frac{-\epsilon_{zp}}{kT}\right) \prod_{j=1}^{3n-6} [1 - \exp(\frac{-h\nu_j}{kT})]^{-1} \qquad (8\text{-}45)$$

where ϵ_{zp} is defined by Eq. (8-42).

8-4.3. The Evaluation of f_{rot}

The solution of the Schrödinger equation for a rigid rotor having three different principal moments of inertia, I_x, I_y, and I_z, involves rather lengthy mathematical manipulations, and, even then, gives very messy expressions for the energy levels [5,6]. The entire treatment reveals only two important results for the present purposes.

The first result is that for most organic molecules the rotational energy levels are closely enough spaced that we may replace the summation over quantum states by integration as we did above for translational motion.

The second important result concerns the effect of molecular symmetry on the occupation of rotational energy states. If a molecule has rotational symmetry, that is, if a molecule has more than one indistinguishable orientation in space, only certain "symmetry allowed" transitions from one rotational state to another are possible. This has the effect of restricting the number of occupiable states available to the molecules. If a molecule has σ indistinguishable orientations in space, then only $1/\sigma$ of the rotational states may be occupied. For example, a dicarboxylic acid, $HOOC(CH_2)_nCOOH$, has $\sigma = 2$ since a rotation of the molecule through 180° produces an orientation which is indistinguishable from the original orientation. The problem of assigning symmetry number σ to a molecule is not always so simple [7,8]. Difficulties are encountered particularly in molecules which have internal rotations or inversions, such as in the case of methylamine.

The result of solution of the Schrödinger equation for the rigid rotor with principal moments of inertia I_x, I_y, and I_z, followed by integration over quantum numbers gives

$$f_{rot} = \frac{8\pi^2(8\pi^3 I_x I_y I_z)^{1/2}(kT)^{3/2}}{\sigma h^3} \qquad (8\text{-}46)$$

8-5. ISOTOPIC SUBSTITUTION IN A MOLECULE [9,10]

Let us now consider the ratio of the "sums over states" defined in Eq. (8-10) for a normal molecule and that of the same molecule in which we have substituted one atom with an isotope. We let A be the normal molecule and A' be the isotopically substituted molecule. We have

$$\frac{\displaystyle\sum_i^A \exp(-\epsilon_i/kT)}{\displaystyle\sum_i^{A'} \exp(-\epsilon_i'/kT)} =$$

$$\frac{\exp(-\epsilon_{zp}/kT)\displaystyle\prod_{j=1}^{3n-6}[1-\exp(-h\nu_j'/kT)]m_A^{3/2}\,\sigma_{A'}[I_x I_y I_z]^{1/2}}{\exp(-\epsilon_{zp}'/kT)\displaystyle\prod_{j=1}^{3n-6}[1-\exp(-h\nu_j/kT)]m_{A'}^{3/2}\,\sigma_A[I_x' I_y' I_z']^{1/2}} \tag{8-47}$$

by combination of Eqs. (8-10), (8-21), (8-42), (8-45), and (8-46). All of the mass independent constants have been cancelled out in Eq. (8-47), and we have used the fact that isotopic substitution does not change the electronic energy to eliminate that term. The primed quantities in Eq. (8-47) refer to the isotopically substituted molecule.

Equation (8-47) can be written entirely in terms of vibrational quantities if we make use of the Redlich-Teller product rule [11]:

$$\left[\frac{m_A}{m_{A'}}\right]^{3/2}\left[\frac{I_x I_y I_z}{I_x' I_y' I_z'}\right]^{1/2} = \left[\frac{m_i}{m_i'}\right]^{3/2}\prod_{j=1}^{3n-6}\frac{\nu_j}{\nu_j'} \tag{8-48}$$

where m_A and $m_{A'}$ are the molecular masses of the normal and substituted molecule, respectively, and m_i and m_i' are the masses of the normal and isotopic atom substituted. This equation is valid for the case of isotopic substitution under consideration if the vibrations are harmonic and the rotational and vibrational degrees of freedom are independent; both cases are approximations which we have already made in arriving at Eq. (8-47).

Substitution of Eq. (8-48) into (8-47) gives

$$\frac{\displaystyle\sum_i^A \exp(-\epsilon_i/kT)}{\displaystyle\sum_i^{A'} \exp(-\epsilon_i'/kT)} = \exp\left(\frac{-\Delta\epsilon}{kT}\right)\frac{m_i^{3/2}\sigma_{A'}\displaystyle\prod_{j=1}^{3n-6}\nu_j[1-\exp(-h\nu_j'/kT)]}{m_i'^{3/2}\sigma_A\displaystyle\prod_{j=1}^{3n-6}\nu_j'[1-\exp(-h\nu_j/kT)]}$$

$$\tag{8-49}$$

where we have defined

$$\Delta\epsilon = \epsilon_{zp} - \epsilon'_{zp} = \sum_{j=1}^{3n-6} h\left(\frac{\nu_j - \nu'_j}{2}\right) \tag{8-50}$$

8-5.1. Isotope Effects on Equilibria

When we speak of the isotope effect on an equilibrium, we actually wish to know the ratio of the equilibrium constants for the reaction of some normal molecule A and for the reaction of the corresponding isotopically substituted molecule A':

$$A + R \xrightleftharpoons{K} B + P$$

$$A' + R \xrightleftharpoons{K'} B' + P$$

Since the isotope must end up in one of the products, one of the products, B' in the example here, also involves isotopic substitution.

The ratio of equilibrium constants, K/K', for these reactions is, of course, just the equilibrium constant for the isotope exchange reaction,

$$A + B' \xrightleftharpoons{K_{ex}} A' + B$$

where

$$K_{ex} = K/K' \tag{8-51}$$

Assuming that the symmetry numbers cancel, and noting that the isotope mass ratio must cancel in the exchange, we substitute Eq. (8-49) into Eq. (8-9) to obtain

$$K_{ex} = \frac{\exp\left(-\Delta\epsilon_B/kT\right) \displaystyle\prod_{j=1}^{3n_B-6} \nu_j[1 - \exp(-h\nu'_j/kT)] \displaystyle\prod_{i=1}^{3n_A-6} \nu'_i[1 - \exp(-h\nu_i/kT)]}{\exp\left(-\Delta\epsilon_A/kT\right) \displaystyle\prod_{j=1}^{3n_B-6} \nu'_j[1 - \exp(-h\nu_j/kT)] \displaystyle\prod_{i=1}^{3n_A-6} \nu_i[1 - \exp(-h\nu'_i/kT)]}$$

$$\tag{8-52}$$

in which n_A is the number of atoms in molecule A and n_B the number in molecule B. The primed quantities refer to the isotopically substituted molecules.

Quite obviously, for most reactions of organic molecules, many of the terms remaining in Eq. (8-52) will cancel completely, since those terms involving frequencies which are unaffected by isotopic substitution at the one position will be the same for the normal and isotopically substituted species.

The case of hydrogen isotopes furnishes an important but relatively simple example of the contributions of the various terms in Eq. (8-52) to the value of K_{ex}. These cases are relatively simple because of several factors. First, hydrogen can be considered to participate in only three vibrational modes in a large organic molecule: one stretching vibration and two bending vibrations. The vibrational frequencies are fairly high, almost never being lower than 600 cm^{-1} even for the bending modes. Further, because of the small mass of hydrogen, deuterium, and tritium, the reduced mass for a vibration in which these atoms participate is essentially equal to the mass of the isotope if we are dealing with most organic molecules.

For the isotopes of hydrogen, from Eq. (8-28) we obtain:

$$\nu_j / \nu_j' \cong (m_i' / m_i)^{1/2} \qquad\qquad\qquad (8\text{-}53)$$

which allows us to work with Eq. (8-52) knowing only the frequencies for the unsubstituted molecule. Some typical hydrogen stretching and bending frequencies are shown in Table 8-1.

As a numerical example, let us consider the deuterium isotope effect on the proton transfer equilibrium,

$$ROH + R_3C^- \rightleftharpoons RO^- + R_3CH$$

for which

$$ROH + R_3CD \xrightleftharpoons{K_{ex}} ROD + R_3CH$$

is the pertinent exchange reaction.

We may note immediately that the ratios of frequencies appearing in Eq. (8-52) cancel by the use of Eq. (8-53):

$$\nu_j \nu_i' / \nu_j' \nu_i = 1 \quad \text{for all i and j} \qquad\qquad (8\text{-}54)$$

TABLE 8-1

Approximate Frequencies of Some
X-H Bond Vibrations

Vibration	ν, cm^{-1}	E_{zp}^{a}	$e^{-h\nu/kT}$ (300 K)
R_3C-H Stretch	2900	4.1	9.7×10^{-7}
R_3C-H Bend	~1300	1.8	2.0×10^{-3}
RO-H Stretch	3600	5.1	3.5×10^{-8}
RO-H in-plane bend	~1400	2.0	1.3×10^{-3}
RO-H out-of-plane bend	~1000	1.4	8.5×10^{-3}
R_2N-H Stretch	3500	4.9	5.6×10^{-8}
$R\overset{O}{\underset{}{C}}$-H Stretch	2800	3.9	1.6×10^{-6}
$R\overset{O}{\underset{}{C}}$-H in-plane bend	~1400	2.0	1.3×10^{-3}
$R\overset{O}{\underset{}{C}}$-H out-of-plane bend	~ 900	1.3	1.4×10^{-2}
C=C-H Stretch	3050	4.3	4.8×10^{-7}
C=C-H in-plane bend	1100	1.6	5.3×10^{-3}
C=C-H out-of-plane bend	900	1.3	1.4×10^{-2}
C≡C-H Stretch	3300	4.7	1.4×10^{-7}
C≡C-H bend	625	0.9	5.1×10^{-2}

$^{a}E_{zp} = h\nu/2$, in units of kcal/mole. The values of h and k in various
units are as follows:
  h = 6.547 x 10^{-27} erg-sec = 2.826 x 10^{-3} cm-kcal/mole
  k = 1.371 x 10^{-16} erg-deg^{-1} = 1.974 x 10^{-3} kcal/mole-deg
  h/kT = 4.77 x 10^{-3} cm at 300 K

Also, the product terms reduce to three terms each, involving the stretching and two bending vibrations of the hydrogen or deuterium:

$$K_{ex} = \exp\left[\frac{-(\Delta\epsilon_{R_3CH} - \Delta\epsilon_{ROH})}{kT}\right]$$

$$\frac{\overset{R_3CD}{\underset{j}{\prod}}[1 - \exp(-h\nu_j'/kT)] \overset{ROH}{\underset{i}{\prod}}[1 - \exp(-h\nu_i/kT)]}{\overset{R_3CH}{\underset{j}{\prod}}[1 - \exp(-h\nu_j/kT)] \overset{ROD}{\underset{i}{\prod}}[1 - \exp(-h\nu_i'/kT)]} \tag{8-55}$$

By the further use of Eq. (8-53), we may write

$$\Delta\Delta\epsilon = \Delta\epsilon_{R_3CH} - \Delta\epsilon_{ROH} = \overset{R_3CH}{\sum} h\frac{[1 - (m_H/m_D)^{1/2}]\,\nu}{2}$$

$$- \overset{ROH}{\sum} h\frac{[1 - (m_H/m_D)^{1/2}]\,\nu}{2} \tag{8-56}$$

where the sums include only the three pertinent frequencies for each molecule.

Taking the frequencies listed in Table 8-1, and evaluating the constants in Eq. (8-56), we find

$$\Delta E_0 = 6.023 \times 10^{23}\,\Delta\Delta\epsilon = 4.14 \times 10^{-4}\,(2900 + 1300 + 1300 - 3600 -$$

$$1400 - 1000) = -0.207 \text{ kcal/mole} \tag{8-57}$$

and

$$\exp\left[\frac{-(\Delta\epsilon_{R_3CH} - \Delta\epsilon_{ROH})}{kT}\right] = \exp\left[\frac{0.207}{0.592}\right] = 1.42 \tag{8-58}$$

We must now examine the terms in $\exp(-h\nu/kT)$ appearing in the right-most portion of Eq. (8-55). The values of these terms are as follows:

$$\nu = 2900,\ \exp\left(-\frac{2900h}{kT}\right) = 9.7 \times 10^{-7},\ \exp\left(-\frac{2900h}{\sqrt{2}kT}\right) = 5.6 \times 10^{-5}$$

$$\nu = 1300,\ \exp\left(-\frac{1300h}{kT}\right) = 2.0 \times 10^{-3},\ \exp\left(-\frac{1300h}{\sqrt{2}kT}\right) = 1.2 \times 10^{-2}$$

$$\nu = 3600,\ \exp\left(-\frac{3600h}{kT}\right) = 3.5 \times 10^{-8},\ \exp\left(-\frac{3600h}{\sqrt{2}kT}\right) = 5.3 \times 10^{-6}$$

$$\nu = 1400, \ \exp\left(-\frac{1400h}{kT}\right) = 1.3 \times 10^{-3}, \ \exp\left(-\frac{1400h}{\sqrt{2}\ kT}\right) = 8.9 \times 10^{-3}$$

$$\nu = 1000, \ \exp\left(-\frac{1000h}{kT}\right) = 8.5 \times 10^{-3}, \ \exp\left(-\frac{1000h}{\sqrt{2}\ kT}\right) = 3.4 \times 10^{-2}$$

We see, then, that the quantities $[1 - \exp(-h\nu/kT)]$ are all quite close to 1. To within 1%, we can now write Eq. (8-55) as follows:

$$K_{ex} = 1.42 \ \frac{(0.988)\ (0.988)\ (0.999)\ (0.992)}{(0.998)\ (0.998)\ (0.991)\ (0.966)} = 1.43 \qquad (8\text{-}59)$$

There are several important general points to be noticed in this example. First, the zero-point energy term, Eq. (8-58), accounts for virtually the entire isotope effect. This term will always operate in the direction such that the heavier isotope will preferentially be in the bond with the highest frequencies. In the above example, the equilibrium constant is greater than unity because the deuterium is in the high-frequency O-D bond on the right-hand side of the reaction.

Second, for any vibrational frequencies greater than about 1300 cm^{-1}, the terms in $[1 - \exp(-h\nu/kT)]$, at room temperature, are so close to unity that they may be neglected in the calculation. Virtually all X-H stretching frequencies, and most X-H bending frequencies, then, contribute only to the zero-point energy term in calculations of isotope effects.

It should also be obvious that as a result of Eq. (8-53), the isotope effects for H-D, H-T, and D-T exchanges will be much larger than for any other isotopic exchange which is likely to be encountered, since the mass ratios for other isotopes are generally quite close to unity.

In those cases where the zero-point energy term is dominant in determining H, D, and T isotope effects, it is possible to establish a relationship between H-D effects and H-T effects. From Eq. (8-56), for the deuterium isotope effect on the zero-point energy term $\Delta\Delta\epsilon$ we have

$$\Delta\Delta\epsilon = [1 - 1/\sqrt{2}]h\Delta\nu/2 = 0.293h\Delta\nu/2 \qquad (8\text{-}60)$$

where $\Delta\nu$ is the total change in frequency:

$$\Delta\nu = \overset{R_3CH}{\sum \nu_j} - \overset{ROH}{\sum \nu_i} \qquad (8\text{-}61)$$

For the corresponding H-T exchange, we have $\Delta\Delta\epsilon'$:

$$\Delta\Delta\epsilon' = \frac{(1 - 1/\sqrt{3})h\Delta\nu}{2} = \frac{0.423h\Delta\nu}{2} \qquad (8\text{-}62)$$

If all other terms can be neglected, for the H-D exchange we have

$$K_{ex} = \exp\left(\frac{-0.293h\Delta\nu}{2kT}\right)$$

or

$$(K_{ex})^{1/0.293} = (K_{ex})^{3.413} = \exp\left(\frac{-h\Delta\nu}{2kT}\right) \tag{8-63}$$

and for the corresponding H-T exchange,

$$K'_{ex} = \exp\left(\frac{-0.423h\Delta\nu}{2kT}\right)$$

or

$$(K'_{ex})^{2.366} = \exp\left(\frac{-h\Delta\nu}{2kT}\right) \tag{8-64}$$

Therefore,

$$(K'_{ex})^{2.366} = (K_{ex})^{3.413}$$

or

$$K'_{ex} = (K_{ex})^{1.44} \tag{8-65}$$

Generally, in dealing with isotopes other than H, D, and T, one must use the full Eq. (8-52). Since Eq. (8-53) will generally not be valid for heavy isotopes, and since the vibrational frequencies for bonds involving heavy atoms are much lower than those for hydrogen, terms other than zero-point energy will be important in these cases [9,10]. The principles, however, are the same as those we have considered.

8-5.2. Secondary Kinetic Isotope Effects

The term "secondary kinetic isotope effect" is applied to isotope effects resulting from isotopic substitution in any bond of a molecule which is not broken in the rate-determining step of the reaction. We may write a general scheme for such situations:

$$A \xrightleftharpoons{K^*} [TS] \longrightarrow product$$

$$A' \xrightleftharpoons{K^{*'}} [TS'] \longrightarrow product'$$

where it is understood that the isotopic substitution must not be in a bond which is being made or broken at the transition state. Since

$$k/k' = K^*/K^{*'} \tag{8-66}$$

we may consider k/k' to be the equilibrium constant for the exchange reaction

$$A + [TS'] \rightleftharpoons A' + [TS]$$

which is formally the same as the equilibrium case that we have already considered. Thus, Eq. (8-52) is valid for this exchange if we simply recognize that B is replaced by TS.

If one has a model for a transition state for a particular reaction which allows the estimation of the frequencies of the vibrations at the transition state, then Eq. (8-52) can be used to calculate the secondary kinetic isotope effects on the reaction. Conversely, measurement of secondary kinetic isotope effects can greatly aid in formulating a model for the transition state for a reaction.

In an earlier chapter, we discussed some secondary kinetic deuterium isotope effects on solvolysis reactions, and found that k_H/k_D was generally greater than 1, ranging from about 1.15 to 1.23 for S_N1 type solvolyses, when the deuterium substitution was made on the α-carbon. The pertinent exchange reaction for these situations is

$$\begin{matrix} R \\ R \end{matrix} H\text{-}C\text{-}X + \left[\begin{matrix} R \\ C^{\delta+}X^{\delta-} \\ D' \\ R \end{matrix} \right] \rightleftharpoons \begin{matrix} R \\ R \end{matrix} D\text{-}C\text{-}X + \left[\begin{matrix} R \\ C^{\delta+}X^{\delta-} \\ H' \\ R \end{matrix} \right]$$

For the saturated C-H frequencies, we may use those listed in Table 8-1, although it should be realized that these will depend somewhat on the identity of X. At the transition state for the reaction, we expect that the carbon has become nearly sp^2 hybridized, and that it has substantial positive charge. An aldehyde C-H might be expected to be quite similar to the C-H at the transition state since it has partial positive charge on the carbon and is sp^2 hybridized [12]. Using the aldehyde frequencies listed in Table 8-1, we then have

$$\Delta\Delta\epsilon = (h/2) \, (1 - 1/\sqrt{2}) \, (2800 + 1400 + 900 - 2900 - 2 \times 1300)$$

and

$$\exp\left(\frac{-\Delta\Delta\epsilon}{kT}\right) = \exp\left(\frac{0.293h400}{2kT}\right) = 1.32 \tag{8-67}$$

Inclusion of the other terms in Eq. (8-52) gives

$$k_H/k_D = 1.32 \frac{(1.00)\ (0.998)^2\ (0.991)\ (0.951)\ (1.00)}{(1.00)\ (0.988)^2\ (0.999)\ (0.986)\ (1.00)} = 1.29 \tag{8-68}$$

which is in reasonable agreement with the experimental values for this type of reaction. Either a slight decrease in the frequencies of the reactant or a slight increase in those for the transition state would lead to even better agreement.

It is worth noting here that the largest contributor to the isotope effect in this last example is the transformation of a fairly rigid C-H bend of the saturated reactant to a fairly floppy out-of-plane bend for the sp^2 transition state. The decrease in bending frequencies on going from sp^3 to sp^2 to sp hybridized C-H bonds is evident in Table 8-1. We may expect, then, that for any reaction in which the amount of p-character of a C-H bond decreases on going from reactant to transition state, we will find k_H/k_D greater than 1; while for a case in which the p-character of the bond increases, we will find k_H/k_D less than 1. Thus, the secondary deuterium isotope effect resembles what we expect of a steric effect if we consider hydrogen to be larger than deuterium.

In the reverse Diels-Alder reaction [13]

where deuterium substitution is made at the hydrogens marked with asterisks, k_H/k_D is found experimentally to be 1.15, in accordance with expectations from consideration of bending frequencies.

In S_N1 solvolysis reactions, deuterium substitution at the β-carbon also leads to k_H/k_D greater than 1. For example, the solvolysis of gives $k_H/k_D = 1.40$. Such effects are most easily rationalized by assuming that hyperconjugation in the cation

decreases the frequencies of the C–H bond from those in the reactant. In fact, when the β-carbon is prohibited from participating in hyperconjugation, the deuterium isotope effect becomes extremely small. For example, $k_H/k_D = 0.99$ for the solvolysis of the following compound [14],

where hyperconjugation would place a partial double bond at the bridgehead.

The secondary kinetic isotope effects that we have considered above are completely different from the isotope effects occurring in those cases where the bond to the isotope is being made or broken at the transition state. In these cases, we must consider the special nature of the vibration of the transition state which converts transition state into reactant or product. We shall see in the next chapter that transition states are quite similar to normal molecules, but differ in the important respect that one of the vibrations of the transition state is very special. We will see that we can still use Eq. (8–52) to calculate isotope effects on rates, but that it must be slightly modified to take account of this special vibration.

PROBLEMS

1. Derive expressions for $\Delta G°$, $\Delta H°$, and $\Delta S°$ in terms of electronic energies, zero-point energies, and f's, for the reaction

$$A + B \rightleftharpoons C + D$$

2. If a particle is constrained to move on a circle of radius r, and the potential energy of the particle is constant for any point on the circle, the Hamiltonian operator for the particle is

$$H = \frac{-[\hbar^2/2r^2 m] \partial^2}{\partial \phi^2}$$

where r and ϕ are the polar coordinates. Find the eigenfunctions and eigenvalues of H. Show that the results give the (4n+2) π-electron rule for cyclic conjugated hydrocarbons.

3. Prove the commutative relationship, $H_j L - LH_j = -h\nu_j L$, where H_j is defined in Eq. (8-23), and $L = \partial/\partial x_j + 2\alpha x_j$. Knowing that the eigenvalues of H_j can never be less than 0, prove that $\epsilon_n = (n + 1/2)h\nu_j$ are the only possible eigenvalues for H_j.

   HINT: Show that there is some lowest eigenfunction of H_j, B, which is annihilated by L; that is, that LB = 0. Then solve for B.

4. Calculate the deuterium isotope effect on the equilibrium

$$
\begin{array}{c}
\text{COOH} \\
/ \\
\text{CH}_2 \\
\backslash \\
\text{COOH}
\end{array}
\; + \text{B} \rightleftharpoons
\begin{array}{c}
\text{COOH} \\
/ \\
\text{CH}_2 \\
\backslash \\
\text{COO}^-
\end{array}
\; + \text{BH}^+
$$

where only the reacting H is substituted by D. Assume the following frequencies:

   COO-H stretch, 3600 cm^{-1}; in-plane bend, 1300 cm^{-1}; and out-of-plane bend, 600 cm^{-1}.
   BH^+ stretch, 3400; doubly degenerate bend, 1400 cm^{-1}.

5. Devise an experiment using secondary isotope effects to find out whether the two bonds are formed simultaneously or consecutively in a Diels-Alder reaction:

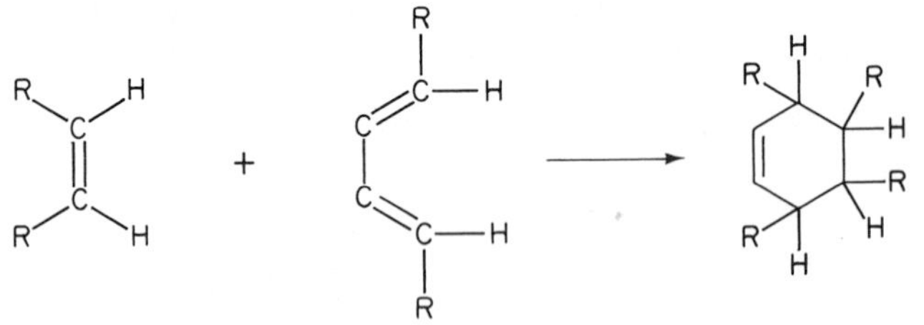

REFERENCES

1. L. P. Hammett, <u>Physical Organic Chemistry</u>, 1st ed. , McGraw-Hill Book Co. , New York, New York, 1940, pp. 73ff. Herein is given a development closely along the same lines.

2. For a more general treatment of equilibria, see H. Eyring, D. Henderson, B. J. Stover, and E. M. Eyring, Statistical Mechanics and Dynamics, John Wiley and Sons, Inc., New York, New York, 1964, pp. 19ff; F. C. Andrews, Equilibrium Statistical Mechanics, John Wiley and Sons, Inc., New York, New York, 1963.

3. T. L. Hill, Introduction to Statistical Thermodynamics, Addison-Wesley Publishing Co., Reading, Massachusetts, 1960. This volume presents the derivation of distribution functions from more basic concepts.

4. W. J. Moore, Physical Chemistry, 2nd ed., Prentice-Hall, Inc., Englewood Cliffs, New Jersey, 1955, Chap. 12. A good elementary discussion is presented.

5. L. F. Phillips, Basic Quantum Chemistry, John Wiley and Sons, Inc., New York, New York, 1965, Chap. 2. This chapter contains a more complete development.

6. L. Pauling and E. B. Wilson, Introduction to Quantum Mechanics, McGraw-Hill Book Co., New York, New York, 1935, pp. 275ff.

7. S. W. Benson and J. H. Buss, J. Chem. Phys., 29, 546 (1958).

8. F. C. Andrews, Equilibrium Statistical Mechanics, John Wiley and Sons, Inc., New York, New York, 1963, p. 145.

9. L. Melander, Isotope Effects on Reaction Rates, The Ronald Press Co., New York, New York, 1960.

10. R. P. Bell, The Proton in Chemistry, 2nd ed., Cornell University Press, Ithaca, New York, 1973, Chap. 11.

11. O. Redlich, Z. physik. Chem., B28, 371 (1935).

12. A. Streitwieser, R. H. Jagow, R. C. Fahey, and S. Suzuki, J. Amer. Chem. Soc., 80, 2326 (1958).

13. S. Seltzer, Tetrahedron Letters, 1962, 457.

14. V. J. Shiner, J. Amer. Chem. Soc., 82, 2655 (1960).

Chapter 9

TRANSITION STATE THEORY AND PRIMARY ISOTOPE EFFECTS

9-1. POTENTIAL ENERGY SURFACES

In dealing with chemical reactions under ordinary circumstances, we are considering systems in which the potential energy depends only on the distances between the atoms of which the system is composed. For two atoms, the potential energy is dependent only on the distance of separation. In general, such a system will involve both attractive and repulsive interactions, but we shall speak of a system being attractive if a stable diatomic molecule is formed and repulsive if no stable molecule can form. The two situations are diagrammed in Figure 9-1, where potential energy is shown as a function of the distance between the two atoms.

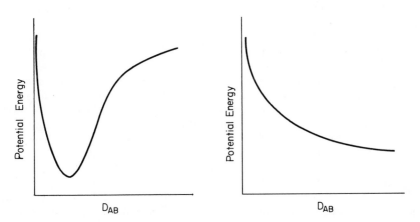

Figure 9-1. Diagrams showing potential energy as a function of distance of separation for a two atom system. The diagram on the left shows an attractive system which forms a diatomic molecule. The diagram on the right shows a repulsive system.

For a system involving three or more atoms, we must specify 3n - 6 positional coordinates, where n is the number of atoms in the system, in order to know the location of each atom relative to all of the others. In such a system, then, potential energy is a function of 3n - 6 variables, and it becomes impossible to make a two dimensional drawing showing potential energy as a function of all coordinates simultaneously. We may, however, use the device of a contour diagram to represent potential energy as a function of any two coordinates simultaneously.

A three-atom system, A, B, C, requires three positional coordinates in order to determine the potential energy. We may hold any one of the coordinates fixed at a given value, and diagram the potential energy as a function of the two remaining coordinates. Some possible contour diagrams for a three atom system are shown in Figure 9-2. Figure 9-2a shows a minimum in the potential energy at fairly small values of d_{AB} and large values of d_{BC} indicating that the system can form a stable AB molecule. At the large values of d_{BC} the potential energy varies with d_{AB} in just the manner shown in Figure 9-1a for a two-atom system. At large values of d_{AB}, Figure 9-2a shows the potential energy varying with d_{BC} in the manner shown in Figure 9-1b for a two-atom system. Thus, Figure 9-2a represents a system in which a diatomic AB molecule exists which has only repulsive interactions with the atom C. Figure 9-2b represents a system in which no stable molecules are formed, and Figure 9-2c represents a system in which a stable triatomic molecule can be formed.

Figure 9-2d represents the type of system with which we shall be concerned in our discussion of the theory of reaction rates. In this system, two stable diatomic molecules can exist, both AB and BC, and each molecule has repulsive interactions with the third atom. Movement of a system from the region at the lower right of this diagram to the region at the upper left represents the reaction of the molecule AB with atom C to form the molecule BC with the atom A.

Before proceeding with our discussion of potential energy surfaces such as that depicted in Figure 9-2d, however, we must pay some attention to the behavior of the potential energy of the system with respect to the third coordinate. It is easily conceivable that a triatomic system could give a contour diagram like that of Figure 9-2c for one given value of ϕ, and like that of Figure 9-2b for another value of ϕ. There is an alternative way of constructing such potential energy contour diagrams that is more useful for the purpose of discussion of reactions. This method involves setting the third coordinate at the value which gives the lowest potential energy for any given values of the other two coordinates. The resulting potential energy diagram then shows the minimum potential energy which the system can have for the values of the coordinates considered.

An example of this alternative type of contour diagram is shown in Figure 9-3 for the six-atom CH_5^- system [1]. In this diagram the minimum

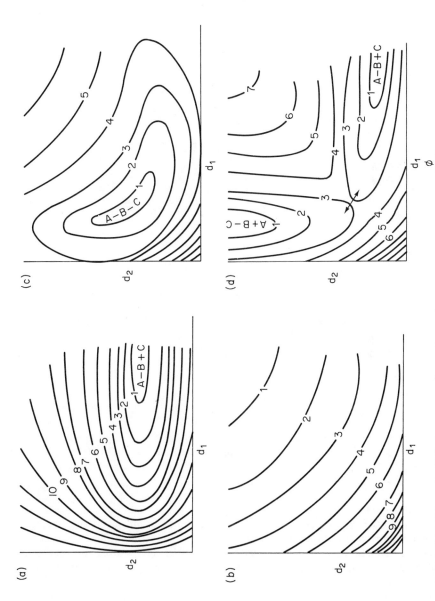

Figure 9-2. Possible potential energy surfaces for a three atom system: A d$_2$ B d$_1$ C with ϕ held constant. Potential energy increases as contour number increases.

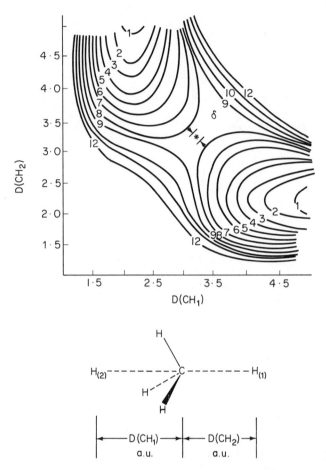

Figure 9-3. Contour diagram showing the minimum potential energy attainable by the CH_5^- system [1] for the depicted values of $D(CH_1)$ and $D(CH_2)$. Contours are at 0.01 a.u. intervals. (0.01 a.u. of energy $\cong$ 6 kcal/mole; 1 a.u. of distance = 0.592 A.)

potential energy of the system for given distances of approach of a hydride ion and separation of an oppositely situated hydrogen from the carbon is shown. The reaction under consideration is obviously that of an S_N2 attack of hydride on methane with another hydride as leaving group. The calculations on this system, which were carried out by the Hartree-Fock MO method, show that the three "nonreacting" hydrogens move from their tetrahedral positions in methane to positions coplanar with the carbon at the positions where the entering and leaving groups are equidistant from the carbon.

The point marked with an asterisk (immediately below δ) in Figure 9-3 is a unique point in the many-dimensional surface. At this point, the potential energy is at a minimum with respect to motion of the system in any direction except one, and with respect to motion in this one direction, the potential energy is at a maximum. Any point on a potential energy surface (many-dimensional) which meets the condition that the potential energy is at a maximum with respect to motion in one direction, and at a minimum with respect to motion in any orthogonal direction, is called a "saddle point." We shall refer to the direction with respect to which the potential energy is at a maximum as the "reaction coordinate" for reasons which will become obvious as we proceed. In Figure 9-3, the saddle point is located at $d(CH_1) = d(CH_2) = 3.30$ a.u., the nonreacting H-C distances at 2.01 a.u., and the other coordinates having values which give the system D_{3h} symmetry. The reaction coordinate is given by

$$V_{rc} = (1/\sqrt{2}) \, [d(CH_1) - d(CH_2)]$$

which is a vector oriented at $-45°$ to the $d(CH_2)$ axis of the figure.

The importance of saddle points in potential energy surfaces arises from the fact that a system passing from one minimum on the surface to another must attain at least the potential energy characteristic of the saddle point connecting the two minima [2]. In Figure 9-3, for example, in order to pass from the minimum at the lower right of the diagram to the minimum at the upper left, the system must attain at least the potential energy at the point marked with the asterisk. In this case, the passage of the system from one minimum to the other corresponds to the S_N2 displacement of hydride ion on methane.

9-2. THE THEORY OF ABSOLUTE REACTION RATES [3]

For a simple general reaction, $A + B \rightleftarrows D + C$, we shall assume that the reactants and products correspond to minima in the potential energy surfaces for the system, and that these minima are connected by a single saddle point, as is the case in Figure 9-3. A reaction occurs any time a system moves from one minimum to the other along any continuous line connecting the two. This motion of a system on the potential energy surface occurs by interconversion of kinetic and potential energies of the system. If we have a large collection of systems at thermal equilibrium, since the number of systems having any given total energy is given by the Boltzmann distribution law, it should be obvious that the most probable paths for conversion of reactants into products will be those which pass through the saddle point. Furthermore, if we deal with systems for which the energy at the saddle point is considerably greater than the average energy of the systems, only a small fraction of the systems will possess enough energy to pass from reactants to products in any given instant of time, and if the kinetic energy of any system can be rapidly redistributed among other systems, once a system has passed through the saddle point, it will continue on to products. We then reach the important conclusion that the rate of passage of systems through the saddle point is a measure of the rate of a reaction.

Transition state theory begins by considering the number of systems having the value for the reaction coordinate which is characteristic of the saddle point, and then considers the average velocity of systems along the reaction coordinate to arrive at the rate of passage of systems from reactants to products.

It follows from our previous discussion of statistical thermodynamics that we can choose any range of configurations of a system that we wish and consider these as a "species." For our present purposes. we define a species TS as those systems whose configurations correspond to a value of the reaction coordinate within a range δ of the saddle point, and any values for the other orthogonal coordinates. One can then define an equilibrium constant

$$K = \frac{[TS]}{[A][B]} \frac{\gamma_*}{\gamma_A \gamma_B} \tag{9-1}$$

where A and B are reactants, just as for any species. The expression for K in terms of the Boltzmann sums over states will be completely analogous to those developed in the last chapter for "ordinary" equilibria. For the species A and B, the evaluation of the sums over states in terms of rotational, translational, and vibrational energy factors is exactly as we have discussed. For the species TS*, however, we must recognize that one of the 3n - 6 vibrations is the motion along the reaction coordinate, which is quite different from the other vibrations. For TS, then, we have only 3n - 7 normal vibrations, and one special type of motion which must be treated in a different manner.

If the range δ to which we restrict values for the reaction coordinate of TS is chosen to be quite small, then, within this range, we may treat the potential energy as a constant. Thus, the motion of the system along the reaction coordinate within the range δ is analogous to the translational motion of a particle in a one-dimensional box of length δ. We may therefore assign an effective mass m* to the system undergoing this motion and write the Boltzmann sum for this translational mode, q^*_{trans}, just as we did in the last chapter for ordinary translational states:

$$q^*_{trans} = (2\pi m*kT)^{1/2} \delta/h \tag{9-2}$$

The equilibrium constant K in Eq. (9-1) then contains the f_{rot}, f_{trans}, and f'_{vib} terms for TS, A, and B just as for an ordinary equilibrium, except that f'_{vib} for TS is

$$f'_{vib}(TS) = q^*_{trans} \prod_{j=1}^{3n*-7} \exp(\frac{-h\nu_j}{2kT}) \; [1-\exp(\frac{-h\nu_j}{kT})]^{-1} \tag{9-3}$$

For convenience in further discussion, we define

$$\epsilon^*_{zp} = \sum_{j=1}^{3n*-7} \frac{h\nu_j}{2} \tag{9-4}$$

and

$$f^*_{vib} = \prod_{j=1}^{3n*-7} [1 - \exp(\frac{-h\nu_j}{kT})]^{-1} \tag{9-5}$$

We may now note that the K in Eq. (9-1) is dependent on δ from Eqs. (9-2) and (9-3). To remove this dependence, we define a new quantity, K*:

$$q^*_{trans} K^* = K$$

or

$$K^* = K/q^*_{trans} \tag{9-6}$$

From Eq. (9-1), we now have

$$[TS] = \frac{q^*_{trans} K^* [A] \; [B] \; \gamma_A \gamma_B}{\gamma_*} \tag{9-7}$$

At thermal equilibrium, the concentration of TS is maintained at the value given by Eq. (9-7) even though systems are constantly passing into and out of the configuration assigned to TS. In order to calculate the rate of formation of products, we now need to determine the rate at which TS is moving through the region δ. Since all the systems are assumed to be at

236   9. TRANSITION STATE THEORY

equilibrium, there will be an equal number of systems moving in the forward and reverse directions along the reaction coordinate. Thus, if we know the average velocity of the systems along the reaction coordinate, the rate of formation of products will be given by

$$\text{rate} = (\text{TS})\ \bar{v}^*/2\delta \tag{9-8}$$

where $\bar{v}^*$ is the average velocity of TS along the reaction coordinate. Starting with the Boltzmann distribution law for velocities,

$$n_v = N_0 \exp(\frac{-m^*v^2}{2kT}) \tag{9-9}$$

which gives the number of systems having a velocity v as a function of the kinetic energy $(1/2)m^*v^2$ we may calculate the average velocity of the systems,

$$\bar{v}^* = \frac{\sum_0^\infty vn_v}{\sum_0^\infty n_v} = \frac{\int_0^\infty v[\exp(-m^*v^2/2kT)]dv}{\int_0^\infty [\exp(-m^*v^2/2kT)]dv} = (2kT/\pi m^*)^{1/2} \tag{9-10}$$

where we have replaced the summation by integration as usual for closely spaced intervals and carried out the integration. The integrals in Eq. (9-10) are in standard form found in any table of standard definite integrals.

Substitution of Eqs. (9-7) and (9-10) into Eq. (9-8) now gives the final result:

$$\text{rate} = \frac{K^* (2\pi m^*kT)^{1/2}\delta/h]\ ([A]\ [B]\ \gamma_A\gamma_B/\gamma_*]\ [2kT/\pi m^*]^{1/2}}{2\delta}$$

$$= \frac{(kT/h)K^*\ [A]\ [B]\ \gamma_A\gamma_B}{\gamma_*} \tag{9-11}$$

which is identical to the expression developed qualitatively in Chapter 2. The present development has emphasized that TS, the transition state, is an ordinary chemical species in all respects except for its motion along the reaction coordinate. Because of this unusual motion, the "equilibrium constant" K^* differs from an ordinary equilibrium constant in that the Boltzmann sums over states for TS* are missing one degree of freedom. In the statistical thermodynamic expression for K^*, we have only $3n^* - 7$ vibrations contributing to both the zero-point energy [Eq. (9-4)] and to the

Boltzmann sums f^*_{vib} [Eq. (9-5)]. We shall now see that this has extremely important consequences for isotope effects on reaction rates.

9-3. PRIMARY KINETIC ISOTOPE EFFECTS [4]

As we have discussed in the previous chapter, the isotope effect on a reaction

$$A + B \rightleftharpoons C$$

can be formulated as the equilibrium constant for an exchange reaction. If we are interested in the rates of reactions, we simply replace C by TS and consider K* for the reaction

$$A + B \xrightarrow{\ K^*\ } TS$$

and the analogous reaction involving isotopically substituted A:

$$A' + B \xrightarrow{\ K'^*\ } TS'$$

Then,

$$k/k' = K^*/K'^* = K^*_{ex} \tag{9-12}$$

where k and k' are the rate constants for the "normal" and isotopically substituted reactions, and K^*_{ex} is the equilibrium constant for the following exchange reaction:

$$A + TS' \xrightarrow{\ K^*_{ex}\ } A' + TS$$

From the general expressions developed in the previous section, along with Eqs. (9-4) and (9-5), we may write K^*_{ex} as

$$K^*_{ex} = \exp\left(\frac{-\Delta\epsilon_{zp}}{kT}\right) \frac{f_{rot(TS)}\, f^*_{vib(TS)}\, f_{trans(TS)}\, f'_{rot(A)}\, f'_{vib(A)}\, f'_{trans(A)}}{f'_{rot(TS)}\, f^{*'}_{vib(TS)}\, f'_{trans(TS)}\, f_{rot(A)}\, f_{vib(A)}\, f_{trans(A)}}$$

$$\tag{9-13}$$

where the primed quantities refer to the isotopically substituted species, and where $\Delta\epsilon_{zp}$ is defined as

$$\Delta \epsilon_{zp} = \sum_{j=1}^{3n_A-6} \frac{h}{2} (\nu'_j - \nu_j) - \sum_{i=1}^{3n^*-7} \frac{h}{2} (\nu'_i - \nu_i) \tag{9-14}$$

Note particularly that the second summation in Eq. (9-14) is over $3n^*-7$ vibrational frequencies and that $f^*_{vib(TS)}$ contains the same $3n^*-7$ modes.

In applying the Redlich-Teller product rules to the ratio of the f's for the transition state, however, we still have $3n^* - 6$ vibrations:

$$\frac{f_{rot(TS)} \, f_{trans(TS)}}{f'_{rot(TS)} \, f'_{trans(TS)}} = (\frac{m_i}{m'_i})^{3/2} \prod_{i=1}^{3n^*-6} \frac{\nu_i}{\nu'_i} \tag{9-15}$$

where m_i is the mass of the atom substituted by the isotope of mass m'_i, and one of the frequencies in the product on the right-hand side of the equation is that for "vibration" of TS along the reaction coordinate. Denoting the vibrational frequency for motion of TS along the reaction coordinate as ν^*, we may use the development detailed in the previous chapter to write

$$K^*_{ex} = \exp\left[\frac{-\Delta \epsilon_{zp}}{kT}\right] \frac{\nu^*}{\nu^{*'}} \prod_{i=1}^{3n^*-7} \frac{\nu_i[1-\exp(-h\nu'_i/kT)]}{\nu'_i[1-\exp(-h\nu_i/kT)]} \prod_{j=1}^{3n_A-6} \frac{\nu'_j[1-\exp(-h\nu_j/kT)]}{\nu_j[1-\exp(-h\nu'_j/kT)]}$$

$$(9-16)$$

The ratio $\nu^*/\nu^{*'}$ in Eq. (9-16) is, to a good approximation, given by

$$\nu^*/\nu^{*'} = (m^{*'}/m^*)^{1/2} \tag{9-17}$$

where m^* is the effective mass along the reaction coordinate and is completely analogous to the reduced mass for an ordinary vibration [4,5].

If the isotopically substituted atom does not move as the system moves along the reaction coordinate, the ratio of frequencies in Eq. (9-17) will be unity, and the noncancelling terms in Eqs. (9-14) and (9-16) will be completely analogous to those for an equilibrium isotope effect as we have already assumed in the previous chapter. Such isotope effects are called secondary kinetic isotope effects.

If the isotopically substituted atom moves as the system moves along the reaction coordinate, the isotope effect on the rate constant for the reaction is called a primary kinetic isotope effect. As an example of this type of isotope effect, let us consider the effect of substituting deuterium for hydrogen in a proton transfer reaction:

$$AH + B \underset{\longleftarrow}{\overset{K^*}{\longrightarrow}} [A \cdots H \cdots B] \longrightarrow A + BH$$

We have, as usual, omitted charges for the sake of generality, and have represented TS as $[A \cdots H \cdots B]$ to emphasize that the hydrogen is moving from A to B at the transition state. The pertinent exchange reaction is

$$AH + [A \cdots D \cdots B] \underset{\longleftarrow}{\overset{K^*_{ex}}{\longrightarrow}} AD + [A \cdots H \cdots B] \qquad (9\text{-}18)$$

We shall assume that the masses of both A and B are much greater than the mass of D so that the reduced masses for those vibrations involving H or D will just be the masses of the isotopes.

In general, the molecule AH will have one A-H stretching and two A-H bending frequencies which we can handle just as we have for the equilibrium situation. The transition state will have four vibrations in which the hydrogen may be involved - a symmetric stretch, an asymmetric stretch, and two bends:

$$\overset{\longleftarrow}{A} \quad \overset{\longrightarrow}{H} \quad \overset{\longleftarrow}{B} \qquad \text{asymmetric stretch, } \nu^*$$

$$\overset{\longleftarrow}{A} \quad \overset{(?)}{H} \quad \overset{\longrightarrow}{B} \qquad \text{symmetric stretch, } \nu_s$$

$$\overset{\uparrow}{A} \quad \overset{\downarrow}{H} \quad \overset{\uparrow}{B} \qquad \text{symmetric bend, } \nu_b$$
$$\otimes \quad \bullet \quad \otimes$$
$$A \quad H \quad B \qquad \text{symmetric bend, } \nu_b$$

$\left. \begin{array}{c} \\ \\ \end{array} \right\}$ assumed to be degenerate

The asymmetric stretching mode is the one which separates A and H while bringing B and H together, or vice versa, and is assumed to be the motion along the reaction coordinate.

The symmetric stretch, as shown, may or may not involve motion of the hydrogen. If the forces between A and H are equal to those between B and H, the hydrogen remains stationary while A and B move symmetrically. If the forces are not equal, the hydrogen will move, at least slightly, along with either A or B. Thus, if the hydrogen is "half-transferred" at the transition state, the reduced mass for the symmetric stretching vibration is insensitive to substitution of deuterium for hydrogen. If the hydrogen is either more or less than half-transferred, then there will be some sensitivity to the isotopic substitution [6], with ν'_s being less than ν_s.

The bending frequencies at the transition state will be related to a reduced mass m_r:

$$m_r = \frac{m_A m_H m_B}{m_A + m_B + m_H} \qquad (9\text{-}19)$$

by the relationship

$$\nu_b/\nu_b' \cong (m_D/m_H)^{1/2} = 2^{1/2} \tag{9-20}$$

just as for the bending frequencies in AH. The bending frequencies at the transition state could conceivably be either greater or less than those in AH. In most cases, however, it is expected that the weakness of bonding at the transition state will result in fairly low bending frequencies.

Finally, the reduced mass for motion along the reaction coordinate in the case where both A and B have masses much greater than that of deuterium is just the mass of the isotope. Equation (9-17) then becomes

$$\nu*/\nu*' = (m_D/m_H)^{1/2} = 2^{1/2} \tag{9-21}$$

In order to gain some feeling for Eq. (9-16), let us take as a first example the case where the A–H stretching frequency is 3300 cm^{-1}, the bending frequencies at the transition state are the same as those of AH, and the symmetric stretching vibration at the transition state is not isotope sensitive. Since exp(-3300h/2kT) is much smaller than unity, Eq. (9-16) for this case becomes simply

$$k_H/k_D = K_{ex}^* = \exp\left[\frac{-h(\nu_{AD} - \nu_{AH})}{2kT}\right] \frac{\nu*}{\nu*'} \frac{\nu_{AD}}{\nu_{AH}}$$

$$= 10.6 \text{ at ca. } 300 \text{ K} \tag{9-22}$$

where ν_{AH} and ν_{AD} are the AH and AD stretching frequencies, respectively. The last two terms cancel, since the reduced masses for both $\nu*$ and ν_{AH} are assumed to be just the isotope masses, and we see that K_{ex}^* is determined completely by the zero-point energy term associated with the AH stretching frequency. Essentially, the zero-point energy difference between AH and AD arising from the stretching vibration has no compensating factor operating at the transition state. Therefore, AH, which has more zero-point energy than AD, is closer in energy to the transition state than is AD.

As a slightly more lengthy example of a calculation of k_H/k_D, let us now consider a case where the following frequencies apply:

For the AH and AD molecules:

stretching frequency: $\nu_1 = 2900$; $\nu_1' = 2045 \text{ cm}^{-1}$

bending frequencies (2): $\nu_2 = \nu_3 = 1300 \text{ cm}^{-1}$

$$\nu_2' = \nu_3' = 920 \text{ cm}^{-1}$$

For the transition states:

symmetric stretch: $\nu_s = 1000 \text{ cm}^{-1}$; $\nu'_s = 900 \text{ cm}^{-1}$

bending frequencies (2): $\nu_{b1} = \nu_{b2} = 600 \text{ cm}^{-1}$

$\nu'_{b1} = \nu'_{b2} = 423 \text{ cm}^{-1}$

These frequencies are consistent with our discussion of reduced masses in that the stretching and bending frequencies of the AH molecule and the bending frequencies of the transition state all show $\nu/\nu' \simeq \sqrt{2}$. The symmetric stretching frequency at the transition state has $\nu/\nu' = 1.1$, somewhat less than if the reduced mass for this motion were simply equal to the mass of the hydrogen or deuterium.

From Eq. (9-14), we have

$$\Delta\epsilon_{zp} = \frac{h}{2}(\nu'_1 - \nu_1 + \nu'_2 - \nu_2 + \nu'_3 - \nu_3) - \frac{h}{2}(\nu'_s - \nu_s + \nu'_{b1} - \nu_{b1} + \nu'_{b2} - \nu_{b2})$$

$$= \frac{h}{2}[(2045 - 2900) + 2(920 - 1300) + (1000 - 900) + 2(600 - 423)]$$

$$= \frac{h}{2} 1161 \text{ cm}^{-1} \tag{9-23}$$

and, at 300°K, $\Delta\epsilon_{zp}/kT = -2.76$. We may then write Eq. (9-16)

$$K^*_{ex} = e^{2.76} \sqrt{2} \frac{2045[1 - \exp^{-2900h/kT}] (920)^2[1 - \exp^{-1300h/kT}]^2}{2900[1 - \exp^{-2045h/kT}] (1300)^2[1 - \exp^{-920h/kT}]^2}$$

$$\times \frac{1000[1 - \exp^{-900h/kT}] (600)^2[1 - \exp^{-423h/kT}]^2}{900[1 - \exp^{-100h/kT}] (423)^2[1 - \exp^{-600h/kT}]^2}$$

$$= 15.8 \frac{(1.00) (0.998)^2 (1.11) (0.986) (0.867)^2}{(1.00) (0.988)^2 (0.992) (0.943)^2}$$

$$= 15.8 \times 0.947 = 15.0 \tag{9-24}$$

Notice particularly that the zero-point energy term alone gives a good estimate of K^*_{ex} even in this more complicated case. This is quite common for H, D, and T isotope effects where the zero-point energy terms are large and the frequencies of most vibrations are quite high. For heavier isotopes, the zero-point energy term will play a less dominant but still quite important role in determining the total isotope effect [4].

Experimentally, k_H/k_D for proton transfer reactions is usually in the range from 5 to 9, although both larger and smaller values are not highly unusual. This normal range of isotope effects corresponds roughly to the loss of just the zero-point energy difference of the AH and AD stretching vibrations on going to the transition states. Larger values of k_H/k_D might result from the bending frequencies of the transition state being considerably smaller than those of AH, as we have seen in the last example. Smaller values of k_H/k_D might result in cases where the proton is only slightly, or nearly completely, transferred at the transition state, making the zero-point energy difference of the symmetric stretching frequencies of the A $\cdots$ H $\cdots$ B and A $\cdots$ D $\cdots$ B transition states quite large.

Very small k_H/k_D values, frequently less than 3, are normally observed in hydride transfer reactions. Such reactions involve the attack of an electrophilic species on an X-H molecule, and can reasonably be formulated to have transition states in which the electrophilic reagent takes advantage of the electron density in the X-H bond to give a nonlinear transition state [7,8]:

$$X\text{-}H + E^+ \rightleftharpoons [X \overset{\cdot\cdot E\cdot\cdot}{\cdot\cdots\cdot}H]^+ \longrightarrow X^+ + EH$$

For such a transition state, the motion along the reaction coordinate may correspond closely to a bending vibration of the reactant X-H. Thus, the zero-point energy "lost" at the transition state will be that of a bending vibration, which is much less than for stretching vibrations.

PROBLEMS

1. Calculate k_H/k_D and k_H/k_T for the reaction

   $$R_3CH + B \rightleftharpoons R_3C^- + BH^+$$

   using the frequencies listed in Table 9-1 and assuming:

   a. The proton is symmetrically located at the transition state and the bending frequencies of the transition state are the same as those of R_3CH.
   b. The bending motions and the symmetrical stretch of the transition state have force constants approximately equal to 0 (i.e., the proton is "free" at the transition state).

2. In the earlier discussion of equilibrium isotope effects, we found that H-D and H-T isotope effects could be related. Under what conditions is the expression $k_H/k_T = (k_H/k_D)^{1.44}$ valid?

3. At very low pressures, many radical or atom combination reactions, such as

$$H + H \longrightarrow H_2$$

are known to take place only at the walls of the container. Why is this true?

4. Show that the isotope effect on the reverse rate constant of a reaction is related to the isotope effect on the forward rate constant through the isotope effect on the equilibrium constant.

5. Show that if no intermediates are involved in a symmetrical reaction

$$A-B + A \rightleftharpoons A + B-H$$

then the transition state must be such that the two A-B distances are equal.

6. Show that if a cyclo-addition reaction, for example,

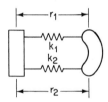

can be modelled by a system with only two force constants:

where k_1 and k_2 depend only on r_1 and r_2, then the transition state for the reaction must have $r_1 \neq r_2$.

    HINT: Show that any point with $r_1 = r_2$ cannot be a saddle point on the potential energy surface. [See J. W. McIver, Jr., Accts. Chem. Res., 7, 72 (1974).]

7. It has been suggested many times that in a series of reactions:

$$A-H + B_i^- \rightleftharpoons A^- + HB_i$$

k_H/k_D should vary with the pK of HB_i and should be at a maximum when $pK_{HB_i} = pK_{AH}$. Why might this behavior be expected? (See R. P. Bell, The Proton in Chemistry, 2nd ed., Cornell University Press, Ithaca, N.Y., 1973, p. 266ff.)

REFERENCES

1. C. D. Ritchie and G. A. Chappell, J. Amer. Chem. Soc., 92, 1819 (1970).

2. J. W. McIver, Jr., Accts. Chem. Res., 7, 72 (1974). This article gives an enlightening discussion of saddle points.

3. H. S. Johnston, Gas Phase Reaction Rate Theory, The Ronald Press, New York, New York, 1966. A considerably more rigorous discussion is presented.

4. L. Melander, Isotope Effects on Reaction Rates, The Ronald Press, New York, New York, 1960.

5. R. P. Bell, The Proton in Chemistry, 2nd ed., Cornell University Press, Ithaca, New York, 1973, pp. 255ff.

6. There is some question as to just how sensitive k_H/k_D is to the position of the proton. See R. P. Bell, The Proton in Chemistry, 2nd ed., Cornell University Press, Ithaca, New York, 1973, pp. 266ff; A. J. Kresge, Chem. Soc. Revs., 2, 475 (1973).

7. M. M. Kreevoy and J. E. C. Hutchins, J. Amer. Chem. Soc., 94, 6371 (1972).

8. G. A. Olah, P. W. Westerman, Y. K. Mo, and G. Klopman, J. Amer. Chem. Soc., 94, 7859 (1972).

Chapter 10

CARBANION CHEMISTRY

10-1. STEREOCHEMISTRY OF CARBANION REACTIONS

The base-catalyzed halogenations of ketones and nitro compounds

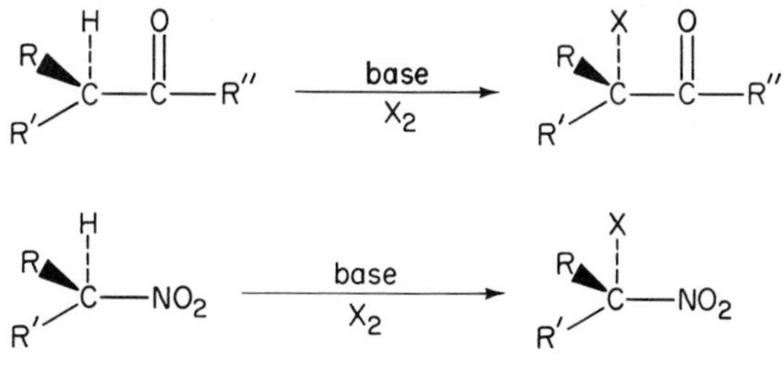

were among the earliest examples of reactions in which carbanions were implicated as intermediates. The kinetics of both of these reactions show general base catalysis, and the rates are independent of halogen concentration under normal conditions. The independence of rate on halogen concentration clearly indicates the existence of an intermediate in the reaction, and the general base catalysis indicates that the formation of this intermediate is the rate-determining step as follows, where A is an activating group such as carbonyl or nitro:

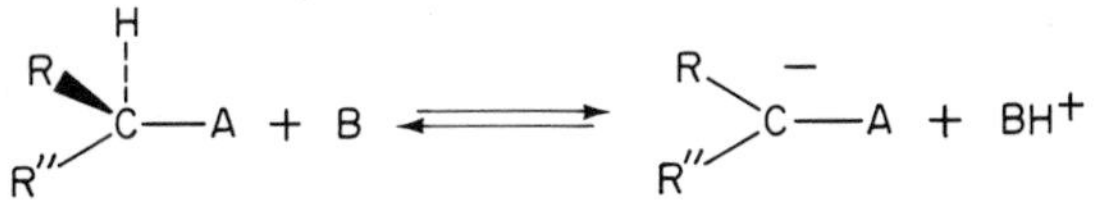

In both of these examples, the intermediate carbanion is expected to have resonance forms which give a trigonal carbon atom, and, therefore, lead to loss of optical activity for reaction of an optically active compound. This, in fact, is observed in most cases of the halogenation reaction. For example, the ketone [1]

undergoes acetate-catalyzed bromination and racemization at equal rates.

Systematic studies of the stereochemistry of other carbanion reactions show, however, a wide variety of behaviors dependent on solvent, counter ion, proton donors in solution, etc. A particularly thorough study of the stereochemistry of reaction I [2]

REACTION I

which proceeds through formation of the alkoxide ion, followed by cleavage of the C-C bond in a reverse condensation

has been reported. Some results are summarized in Table 10-1.

These results may be rationalized by means of a reaction scheme involving ion pairs whose rates of rearrangements, etc., compete with protonation of the carbanion moiety by either solvent or the conjugate acid of the base catalyst:

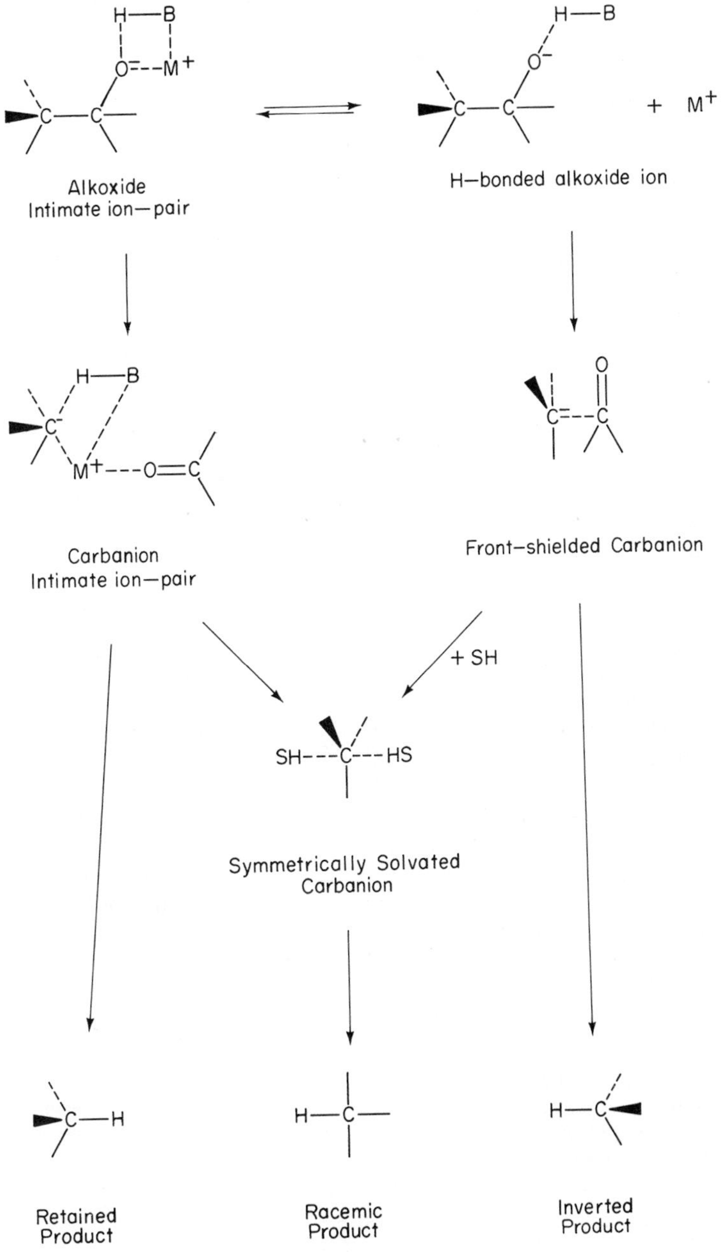
H—B
O⁻--M⁺
Alkoxide
Intimate ion—pair
H—B
O⁻
+ M⁺
H—bonded alkoxide ion
H—B
C⁻
M⁺---O═C
Carbanion
Intimate ion—pair
O
C⁻--C
Front—shielded Carbanion
+ SH
SH---C---HS
Symmetrically Solvated
Carbanion
C—H
Retained
Product
H—C—
Racemic
Product
H—C
Inverted
Product

TABLE 10-1

Stereochemical Course of Reaction I [2]

R	Solvent	Base	Net stereochemistry of 2-phenylbutane
Ethyl	Benzene	$t\text{-BuO}^-K^+$	93% retention
Ethyl	t-Butanol	KOH	93% retention
Ethyl	s-Butanol	$s\text{-BuO}^-K^+$	85% retention
Ethyl	Methanol	MeO^-K^+	11% retention
Ethyl	$O(CH_2CH_2OH)_2$	KOH	23% inversion
Ethyl	$HOCH_2CH_2OH$	KOH	48% inversion
Phenyl	t-Butanol	$t\text{-BuO}^-Li^+$	85% retention
Phenyl	t-Butanol	$R_4N^+OH^-$	100% racemization
Phenyl	DMSO	$t\text{-BuO}^-K^+$	100% racemization

The intimate ion pair route to retained product will be favored by low dielectric constant and poor cation solvating solvents, and by cations which form "tight" ion pairs. The decision between retained or racemic product from the intimate ion pair route will depend upon the ability of BH to donate a proton to the front side of the carbanion in competition with symmetrical solvation of the carbanion. Good dissociating solvents will favor the route through the front-shielded carbanion. The decision between forming inverted product or racemic product will depend upon the rate at which solvent can protonate the shielded intermediate in competition with the rate of symmetrical solvation.

10-2. EXCHANGE AND RACEMIZATION OF HYDROCARBONS

A great deal more information about the competitive processes of ion pair dissociation, protonation, etc., in carbanion reactions has been gained from studies of the stereochemistry of isotope exchange reactions of hydrocarbons:

Under the usual conditions for the study of these reactions, the total pool of hydrogen, from BH^+ and/or the solvent SH, is in great excess of the amount of deuterium introduced in the reactant, and the reaction proceeds to completion to the right.

Starting with an optically active hydrocarbon, two rates may be measured; the rate of formation of RH product and the rate of loss of optical activity. Both rates are generally pseudo-first-order processes and the pseudo-first-order rate constants, k_{ex} and k_α, are defined just as in our earlier discussions of carbonium ion reactions. If each exchange of D for H results in racemization, then $k_{ex} = k_\alpha$. If each exchange occurs with inversion, $k_{ex}/k_\alpha = 0.50$, and if the exchange occurs only with retention, $k_{ex}/k_\alpha = \infty$.

The exchange reactions are most conveniently discussed in terms of a general reaction scheme similar to that above, but involving hydrogen-bonded carbanions rather than ion pairs [3]:

There are several experimental results which strongly indicate that the one hydrogen–bond–donating molecule of either BH or solvent shown in the scheme is all that can be reasonably written. The base–catalyzed tautomerization of the triarylmethane isomer [4]

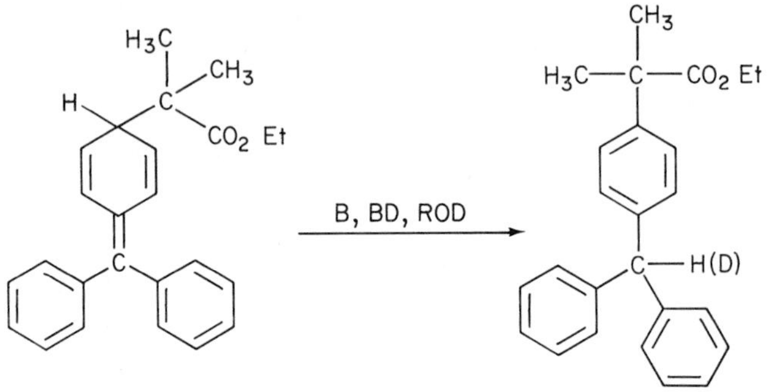

B, BD, ROD

gives only 3% incorporation of deuterium in the product when B is tripropylamine and ROD is deuterated triethylcarbinol. Even with B $= CH_3O^-$ in MeOD, only 50% deuterium is incorporated into the product. Similarly, when the following optically active fluorene [5]

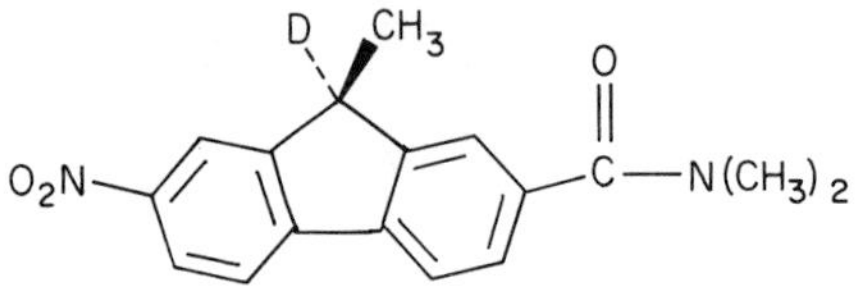

is reacted with tripropylamine and tripropylammonium ion in THF solution with t–BuOH added, the ratio k_{ex}/k_α is found to be 0.1. All of these results can be understood in terms of the scheme with $k_b > k_c$ and k_d, and cannot accomodate an intermediate in which more than one BD or BH molecule are in a position to protonate the carbanion moiety.

If only one BH or BD molecule can be associated with the carbanion, we might also expect that if a primary amine, RNH_2, is reacted with an optically active deuterated hydrocarbon, then by rotation of the ammonium ion we could exchange D for H without racemization. That this actually happens is shown by the fact that k_{ex}/k_α is found to be 56 for the following reaction [6]:

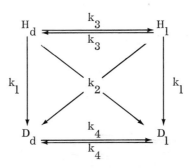

$$X = \overset{\displaystyle O}{\underset{\displaystyle \|}{C}} - N(CH_3)_2$$

It is quite clear in these examples that the result $k_{ex}/k_\alpha > 1$ may be interpreted in terms of the general reaction scheme as showing that $k_c > k_b$ or k_d, and that the result $k_{ex}/k_\alpha < 0.5$ shows that $k_b > k_c$ or k_d. Values of k_{ex}/k_α between 0.5 and 1.0 (i.e., some net inversion) can result from two different possibilities.

Representing the various exchange and stereoconversions by the short-hand scheme,

Scheme I

where H_d represents the dextrorotatory protio-substrate, H_l the levorotatory protio-substrate, D_d the dextrorotatory deuterio-substrate, and D_l

the levorotatory deuterio-substrate, we can easily see the two possibilities for k_{ex}/k_α between 0.5 and 1.0. One possibility would be that $k_2 > k_1$, indicating a preference for backside protonation of the intermediate by solvent. The other possibility is that k_3 is of an appreciable magnitude relative to k_1 and k_2, with $k_1 = k_2$. In the general scheme, the first possibility would correspond to $k_d > k_c$, and the second possibility would correspond to $k_d = k_c$ but with $k_{-a}k_b/(k_{-a} + k_b)$ at least comparable in magnitude to k_d and k_c. That is, the second possibility requires that the protonation of the intermediate be nearly as fast as the exchange of BH for BD in the intermediate.

10-3. KINETICS OF EXCHANGE AND RACEMIZATION REACTIONS

In principle, the simplest method for distinguishing between the two above possibilities would be to stop the reaction after a small fraction of conversion of H_d to D, isolate and resolve the optical isomers, and analyze the levorotatory material for deuterium. If there is an appreciable amount of H_1 in the levorotatory fraction, then k_3 must be comparable to k_1 and k_2. In practice, this method requires a great deal of time and effort in resolving the optical isomers. A purely kinetic method, which is more complex in theory, is actually much easier in practice.

From Scheme I, we can write the kinetic expressions in matrix form:

$$
\begin{bmatrix}
-(k_1+k_2+k_3) & k_3 & 0 & 0 \\
k_3 & -(k_1+k_2+k_3) & 0 & 0 \\
k_2 & k_1 & -k_4 & k_4 \\
k_1 & k_2 & k_4 & -k_4
\end{bmatrix}
\begin{bmatrix}
H_d \\ H_1 \\ D_1 \\ D_d
\end{bmatrix}
=
\begin{bmatrix}
\dot{H}_d \\ \dot{H}_1 \\ \dot{D}_1 \\ \dot{D}_d
\end{bmatrix}
$$

(10-1)

$$KC = \dot{C}$$

You may easily verify that the K matrix is triangularized

$$U^{*}KU = T \tag{10-2}$$

with

$$
U =
\begin{bmatrix}
1/\sqrt{2} & -1/\sqrt{2} & 0 & 0 \\
1/\sqrt{2} & 1/\sqrt{2} & 0 & 0 \\
0 & 0 & 1/\sqrt{2} & -1/\sqrt{2} \\
0 & 0 & 1/\sqrt{2} & 1/\sqrt{2}
\end{bmatrix}
$$

$$T = \begin{bmatrix} -(k_1+k_2) & 0 & 0 & 0 \\ 0 & -(k_1+k_2+2k_3) & 0 & 0 \\ (k_1+k_2) & 0 & 0 & 0 \\ 0 & (k_2-k_1) & & -2k_4 \end{bmatrix}$$

and

$$U*C = C' = 1/\sqrt{2} \begin{bmatrix} H_d + H_1 \\ H_1 - H_d \\ D_d + D_1 \\ D_d - D_1 \end{bmatrix}$$

The resulting equation, $TC' = \dot{C}'$, gives the equations

$$\dot{c}_1' = -(k_1+k_2)c_1' \qquad c_1' = c_0' \exp[-(k_1+k_2)t] \tag{10-3}$$

$$\dot{c}_2' = -(k_1+k_2+2k_3)c_2' \qquad c_2' = -c_0' \exp[-(k_1+k_2+2k_3)t] \tag{10-4}$$

$$\dot{c}_4' = (k_2-k_1)c_2' - 2k_4 c_4'$$

$$c_4' = \frac{-c_0'(k_2-k_1)}{2k_4-k_1-k_2-2k_3} \quad \exp[-(k_1+k_2+2k_3)t] - \exp(-2k_4 t) \tag{10-5}$$

where c_0' is the initial concentration of H_d divided by $\sqrt{2}$.

The forms of the c' vector elements are quite convenient for racemization and exchange kinetics since

$$\alpha = \alpha_0[H_d - H_1 + D_d - D_1] = \alpha_0 \sqrt{2} \ [c_4' - c_2'] \tag{10-6}$$

and

$$\dot{c}_1' = -k_{ex} c_1' \qquad k_{ex} = k_1 + k_2 \tag{10-7}$$

From Eqs. (10-4), (10-5), and (10-6), we have

$$\alpha = \alpha^\circ \left\{ \frac{2(k_2+k_3-k_4)}{k_1+k_2+2k_3-2k_4} \exp[-(k_1+k_2+2k_3)t] \right.$$
$$\left. - \frac{k_2-k_1}{k_1+k_2+2k_3-2k_4} \exp[-2k_4 t] \right\} \tag{10-8}$$

where we have defined $\alpha^\circ = \alpha_0 \sqrt{2} \ c_0'$.

Equation (10-8) shows that the racemization will not follow simple first-order kinetics unless one of the two exponential factors vanishes. That is, the racemization will show first-order kinetic behavior only if $k_2+k_3 = k_4$, or if $k_2 = k_1$. Referring to Scheme I, we can see that (k_2+k_3) can be determined from the initial rate of racemization of H_d, and that k_4 can be obtained from the rate of racemization of D_d under the given reaction conditions. Since $(k_2+k_3)/k_4$ is nothing more than a primary isotope effect on initial rates of racemization, we should expect this ratio to be in the usual range for such isotope effects.

For the reaction of 9-methyl-2-dimethylcarboxamidofluorene [7] in CH_3OD with K^+-OCH_3, $(k_2+k_3)/k_4 = 6.5$ at 25 °C, and the racemization of the H_d follows strict first-order kinetics. Thus, for this case $k_2 = k_1$, and the observed result that $k_{ex}/k_\alpha \simeq 0.92$ must arise from a contribution from the k_3 process of Scheme I. The actual rate constants for Scheme I, found from rates of racemization and exchange treated by Eqs. (10-7) and (10-8), are $k_1 = k_2 = 3.02 \times 10^{-3}$, $k_3 = 0.26 \times 10^{-3}$, and $k_4 = 0.51 \times 10^{-3}$ (in units of M^{-1} sec^{-1}).

10-4. RATES OF PROTON TRANSFER REACTIONS

In our earlier treatment of acid-base catalysis, we have already discussed the fact that proton transfer reactions between most oxygen and nitrogen acids and bases occur at diffusion-controlled rates in the thermodynamically favored direction. That is, for a reaction

$$A\text{-}H + B^- \underset{k_r}{\overset{k_f}{\rightleftharpoons}} A^- + BH$$

where AH and BH are both either nitrogen or oxygen acids, k_f will generally be diffusion controlled if the equilibrium constant for the reaction is greater than 1, and k_r will be diffusion controlled if the equilibrium constant is less than 1. The diffusion rate constant for molecules and ions of moderate size in most solvents is close to 10^{10} M^{-1} sec^{-1}. If either k_f or k_r is diffusion controlled, then the actual rate-determining step in the proton transfer mechanism is the diffusion step and not the actual proton transfer step.

In the proton transfer reactions of many, but not all, carbon acids, both k_f and k_r are found to be much less than diffusion controlled in magnitude. Some representative data for the rates of such reactions are shown in Table 10-2.

It appears that there are two primary factors which affect the rates of proton transfer reactions [8]. First, the transition state for the actual

TABLE 10-2

Rate Constants for Proton Transfer Reactions
in Water at 25 °C [9]

$$HA + B \underset{k_r}{\overset{k_f}{\rightleftharpoons}} A + HB$$

HA	B	k_f (M^{-1} sec^{-1})	k_r (M^{-1} sec^{-1})
H_2O	Phenoxide ion	1.4×10^6	1.4×10^{10}
H_3O^+	Hydroxide ion	1.4×10^{11}	1.0×10^{-3}
H_3O^+	CH_3COO^-	4.5×10^{10}	8.2×10^5
H_3O^+	Triethylamine	2.7×10^{10}	4.0×10^{-1}
H_3O^+	Acetone enolate (on C)	5.0×10^{10}	$\sim 10^{-10}$
H_3O^+	Acetylacetonate (on C)	1.2×10^7	1.2×10^{-3}
H_3O^+	$CH_2NO_2^-$ (on C)	6.8×10^2	4.3×10^{-8}
H_3O^+	$CH(NO_2)_2^-$ (on C)	3.1×10^3	8.0×10^{-1}
H_3O^+	Azulene	1.2	1.3×10^2
H_2O	Acetone enolate (on C)	$\sim 4 \times 10^6$	2.7×10^{-1}
H_2O	$CH_2NO_2^-$ (on C)	2.8×10^{-3}	28.0
H_2O	Trimethylamine	1.3×10^6	2.1×10^{10}

proton transfer step may be stabilized by hydrogen-bonding type interactions between the proton donor and acceptor. As we saw in the discussion of the stabilities of carbanions, effective hydrogen bonding requires that the acceptor (base) have a localized lone pair of electrons. For the hydrogen bonding to stabilize the transition state for the proton transfer, the lone pair of electrons much be localized on the atom to which the proton is being donated. Second, if the transformation of B to BH or of AH to A involves large changes in bond angles or bond lengths within the A or B moieties, the energy required to reach the transition state is increased. This factor is a particular example of the Principle of Least Action which states that those reactions involving the least change in bond angles or lengths have lower activation energies than those involving more change in such geometrical parameters.

The structure of an enolate ion, for example, is such that the negative charge is almost entirely on the oxygen,

$$
\begin{array}{c}
O^- \\
| \\
R \diagdown \overset{C}{} \diagup\kern-0.6em=\kern-0.6em C - R \\
| \\
R
\end{array}
$$

and there is nearly a normal C-C double bond. Protonation on the oxygen to form the enol involves little change in bond lengths and angles and occurs at diffusion controlled rates with acids stronger than the enol. Protonation on carbon, however, involves changing the double to a single C-C bond, and the single to a double C=O bond, along with bond angle changes. Except for very basic enolates and very acidic acids, the protonation on carbon is much below diffusion-controlled rates.

The Principle of Least Action is most easily comprehended by consideration of an identity reaction:

$$
A^- + HA \rightleftharpoons [A \cdots \overset{-}{H} \cdots A]^* \rightleftharpoons AH + A^-
$$

If the potential energy surface for a reaction of this type has only one saddle point separating reactants and products, then the transition state for the reaction must be such that the two A moieties are identical. If the internal geometry of A is different in A^- and HA, then the geometry of A^- must be distorted toward that of HA, and the geometry of HA must be distorted toward that of A^- on going from reactants to transition state. These distortions of geometry involve the stretching and bending of bonds and require the expenditure of energy. The greater the change in geometry, the greater the energy expenditure necessary to reach the transition state [9].

This "reorganization energy" to distort the geometries of the reactants will be compensated by any favorable interactions of HA with A^-, such as hydrogen bonding, at the transition state. If A^- contains a lone pair of electrons localized at the atom to which the proton is being transferred, the hydrogen bonding of HA to A^- may be favorable enough completely to compensate the reorganization energy. For example, the identity reaction of p-nitrophenol with p-nitrophenoxide occurs at very nearly a diffusion-controlled rate, even though the bond lengths and angles of the p-nitrophenoxide ion are considerably different from those of p-nitrophenol because of the resonance form:

In this case, the phenoxide oxygen has three lone pairs of electrons, only one of which is appreciably delocalized.

Since a carbanion has only one lone pair of electrons, any delocalization will decrease hydrogen bonding interactions and will also generally lead to differences in geometry of the carbanion and its conjugate acid. For example, the identity reaction [10]

shows a rate constant of 0.5 M^{-1} sec^{-1} in dimethylsulfoxide solution at 25°C. This amounts to ca. 17 kcal/mole for ΔG^*.

If we consider the unsymmetrical reaction of an acid with an anion whose conversion to conjugate acid involves geometric changes, we must take into account the possibly unsymmetrical nature of the transition state. If the acid is considerably stronger than the conjugate acid of the anion, it seems reasonable to expect that the proton can be transferred with only little reorganization of the anion. If the acid were extremely strong, it could donate a proton to the anion with no prior reorganization and still lower the total energy of the system. Thus, we expect that for a given anion, the structure of the transition state will be productlike for weak acids, and reactantlike for strong acids [9].

In our earlier discussion of the Brønsted catalysis law, we observed that a Brønsted slope different from 0 or 1 can be observed only when the actual proton transfer step of an acid–base reaction is rate determining. It should now be obvious that Brønsted slopes different from 0 or 1 for reactions such as

$$A^- + HB_i \xrightarrow{K_i} AH + B_i^-$$

will be observed over wider variations of K_i for cases where the conversion of A^- to HA involves the greater geometrical changes, and where A^-

is the poorer hydrogen bond acceptor. In the protonation of acetylacetonate ion on carbon [11], for example, the Brønsted slope varies gently from 0.74 to 0.52 as K_i varies from 10^{-5} to 10^{+7}. For normal oxygen or nitrogen bases, the Brønsted slope varies from 1 to 0 over a range of ca. 10^3 in K_i.

10-5. THE RATE-DETERMINING STEP IN EXCHANGE REACTIONS

Many carbon acids are too weak to allow direct observation of the carbanion species in proton transfer reactions. As we have already discussed in connection with stereochemistry, however, it is usually possible to study base-catalyzed isotope exchange reactions, and to obtain from these some information on the rates of the proton transfer reactions of the carbanion intermediates. There have been a large number of studies of these isotope exchange reactions over the past thirty years. Initially, many of these studies had the purpose of attempting to establish the relative acidities of the hydrocarbons by postulating a linear relationship between log k_{ex} and log K_a for a series of compounds studied under the same conditions. From the above discussion, it should be clear that such a postulated relationship is generally false. For example, the fact that phenylacetylene-d_1 undergoes D-H exchange in methanol faster than does fluorene-9-d_1 cannot be taken as evidence that phenylacetylene is a stronger acid than fluorene. Since the phenylacetylide ion is a localized anion, its rate of reaction with acids stronger than phenylacetylene is expected to be diffusion controlled. The same is not true for the fluorenyl anion for the reasons that we have discussed, and in this case, the actual proton transfer step is quite likely to be rate determining.

The identification of the rate-determining step for isotope exchange reactions can be made by the use of kinetic isotope effect studies is some cases. These studies make use of the fact that hydrogen has three isotopes, and that k_H/k_D and k_H/k_T or k_D/k_T can be related to one another by use of the reduced mass relationships discussed in the chapters on isotope effects. The isotope effect, k_D/k_T, is determined by measuring the rate of isotope exchanges of R-D and of R-T in a solvent such as CH_3OH. The ratio k_H/k_T is then determined by measuring the rate of exchange of R-H and R-T in CH_3OD. The experimentally observed ratios are compared with the theoretical relationship between the two ratios, and discrepancies may be attributed to non-rate-determining proton transfer, as we shall now develop [12].

We may write the mechanism for the exchange of R-D in CH_3OH catalyzed by CH_3O^- as follows:

$$R\text{-}D + CH_3O^- \underset{k_{-1}^D}{\overset{k_1^D}{\rightleftharpoons}} [R^- \, DOCH_3] \qquad\qquad (10\text{-}9)$$

$$[R^- \ DOCH_3] + CH_3OH \xrightarrow{\ k_2^D\ } [R^- \ HOCH_3] + CH_3OD \qquad RH \qquad (10\text{-}10)$$

The second step of the reaction may be considered nonreversible since the concentration of CH_3OH is always much greater than that of CH_3OD. For the hydrocarbons under consideration, we may apply the Bodenstein approximation to the above scheme to get the following rate expression:

$$d[RH]/dt = k_2^D \ [R^- \ DOCH_3] = \frac{k_1^D k_2^D}{k_{-1}^D + k_2^D} \ [RD] \ [CH_3O^-] \qquad (10\text{-}11)$$

The observed second-order rate constant k_{obs}^D is then

$$k_{obs}^D = \frac{k_1^D k_2^D}{(k_{-1}^D + k_2^D)} \qquad (10\text{-}12)$$

Analogous expressions are obtained for the observed rate constants for RT exchange, k_{obs}^T, and for RH exchange, k_{obs}^H, which may be combined to give:

$$\frac{k_{obs}^T}{k_{obs}^H} = \frac{k_1^T (k_{-1}^H + k_2^H)}{k_1^H (k_{-1}^T + k_2^T)} \qquad (10\text{-}13)$$

$$\frac{k_{obs}^D}{k_{obs}^T} = \frac{k_1^D (k_{-1}^T + k_2^T)}{k_1^T (k_{-1}^D + k_2^D)} \qquad (10\text{-}14)$$

The key assumption that we make in using these measured isotope effects to obtain k_{-1}/k_2 is that k_2 is independent of isotope. This assumption is reasonable since step 2 in the mechanism is essentially a diffusion controlled step. We shall, then, omit the superscript on k_2 in the subsequent equations.

The definition of several quantities will help simplify the algebraic manipulations of the equations:

$$a^T = k_{-1}^T/k_2 \qquad (10\text{-}15)$$

$$K_T = k_1^T/k_{-1}^T \qquad K_D = k_1^D/k_{-1}^D \qquad K_H = k_1^H/k_{-1}^H \qquad (10\text{-}16)$$

$$K'_T = K_T/K_H \qquad K''_T = K_T/K_D \tag{10-17}$$

Rearrangement of Eq. (10-13) then gives

$$\frac{k^T_{obs}}{k^H_{obs}} = \frac{k^H_{-1}/k^H_1 + k_2/k^H_1}{k^T_{-1}/k^T_1 + k_2/k^T_1} = \frac{1/K_H + k_2/k^H_1}{1/K_T + k_2/k^T_1}$$

$$= \frac{K_T/K_H + (k_2/k^H_1)\,(k^T_1/k^T_{-1})}{1 + (k_2/k^T_1)\,(k^T_1/k^T_{-1})}$$

$$= \frac{K'_T + (k^T_1/k^H_1)\,(1/a^T)}{1 + 1/a^T} = \frac{K'_T a^T + k^T_1/k^H_1}{1 + a^T}$$

$$\therefore \quad \frac{k^T_1}{k^H_1} = \frac{k^T_{obs}}{k^H_{obs}} - a^T\,(K'_T - \frac{k^T_{obs}}{k^H_{obs}}) \tag{10-18}$$

Completely analogous manipulation of Eq. (10-14) gives

$$\frac{k^T_1}{k^D_1} = \frac{k^T_{obs}}{k^D_{obs}} - a^T\,(K''_T - \frac{k^T_{obs}}{k^D_{obs}}) \tag{10-19}$$

Since we have assumed that k_2 is independent of isotope, the quantity K'_T defined in Eq. (10-17) is just the equilibrium constant for the reaction

$$R\text{-}T + CH_3OH \overset{K'_T}{\rightleftharpoons} R\text{-}H + CH_3OT$$

which can be directly measured by equilibrating R-H in partially tritiated methanol. For R-H = 9-methylfluorene, K'_T was measured at several temperatures with the following results: $T = 0\,°C$, $K'_T = 1.35$; $T = 25\,°C$, $K'_T = 1.32$; $T = 50\,°C$, $K'_T = 1.28$; $T = 100\,°C$, $K'_T = 1.23$ [12].

The whole purpose of putting Eqs. (10-18) and (10-19) into the form given is that the isotope effects for a single step can be related by theory to reduced masses. In our earlier discussion, we assumed that the reduced masses for all vibrations of the isotopes were simply the masses of the isotopes and arrived at the expression:

$$k_H/k_T = (k_D/k_T)^{3.25} \tag{10-20}$$

A more sophisticated calculation for hydrocarbons such as 9-methylfluor-ene, allowing for the fact that the reduced masses are not exactly equal to the isotope masses, gives [12]

$$k_H/k_T = (k_D/k_T)^{3.344} \tag{10-21}$$

Equation (10-21) is valid for either equilibrium isotope effects or for isotope effects on the rate constants for an elementary step of a reaction. We may therefore use it to calculate K_T'' from K_T', and the relationship between Eqs. (10-18) and (10-19):

$$K_T'' = K_T'^{\,3.344} \tag{10-22}$$

$$k_1^H/k_1^T = (k_1^D/k_1^T)^{3.344} \tag{10-23}$$

Combination of Eqs. (10-18), (10-19), (10-22), and (10-23) then allows the evaluation of a^T from the measured values of k_{obs}^T/k_{obs}^H, k_{obs}^T/k_{obs}^D, and K_T'.

The experimentally observed isotope effects and the calculated values for a^T are shown in Table 10-3 for three hydrocarbons.

TABLE 10-3

Isotope Effects in Isotope Exchange Reactions
Catalyzed by Methoxide Ion in Methanol [12]

	$(C_6H_5)_3CH$ (97.7°C)	9-Phenyl-fluorene (25°C)	9-Methyl-fluorene (45°C)
k_{obs}^D/k_{obs}^T	1.34	2.50	2.30
k_{obs}^H/k_{obs}^T	1.77	16.0	11.9
k_{-1}^T/k_2	0.66	0.016	0.024
k_1^D/k_1^T	1.85	2.54	2.38

The small value for a^T for the fluorenes show that the actual proton transfer step is rate determining for these hydrocarbons. In the case of triphenylmethane, neither step 1 nor step 2 is completely rate determining. The proton transfer from methanol to the carbanion occurs at nearly the same rate that the initially formed methanol molecule exchanges with solvent methanol. This result is nicely consistent with the result, which we discussed earlier, that the base-catalyzed tautomerization of a cyclohexadiene tautomer of triphenylmethane occurs in methanol solution with approximately 50% exchange with solvent.

10-6. KINETICS OF THE REACTIONS OF ION PAIRS

We have seen in the above discussions that carbanion ion pairs exist in solvents of low dielectric constant, and that the natures of the ion pairs affect the stereochemistry of reactions involving carbanion-like intermediates. These ion pairs also have different reactivities from one another and from free carbanions which affect the kinetics of reactions in which they are involved.

Perhaps the most thorough study of the kinetics of a reaction involving ion pair intermediates is that for the anionic polymerization of styrene in tetrahydrofuran solution [13]. These reactions involve the propagation of a "living polymer" chain, $RCH_2CH\Phi$, by reaction with styrene:

We may write the general reaction scheme for this reaction as follows:

$$[R^-M^+] \underset{\longleftarrow}{\overset{K_d}{\longrightarrow}} R^- + M^+$$

$$[R^-M^+] + styrene \xrightarrow{k_{ip}} [R^-M^+]$$

$$R^- + styrene \xrightarrow{k_{ion}} R^-$$

from which the kinetic expression is easily derived:

$$\frac{d[styrene]}{dt} = -(k_{ip}[R^-M^+] + k_{ion}[R^-]) \, [styrene] \qquad (10\text{-}24)$$

The relative proportions of ion pair and free ion can be varied by either changing the total concentration of the two species, or by addition of a salt, $M^+BF_4^-$, to the solution. Let us define

$$[X] = [R^-M^+] + [R^-] \qquad (10\text{-}25)$$

We have already seen in chapter 6 of this book that K_d in THF solution is quite small. We may therefore safely assume that

$$[R^-] << [R^-M^+] \qquad (10\text{-}26)$$

In the absence of added salt, we have the relationship

$$[R^-] = (K_d/[X])^{1/2} \qquad (10\text{-}27)$$

and substitution into Eq. (10-24) gives the expression

$$k_\Psi = \left[k_{ip} + k_{ion}\sqrt{K_d/[X]} \right][X] \qquad (10\text{-}28)$$

for the pseudo-first-order rate constant k_Ψ.

If $[M^+]$ is varied by addition of salt to the solution, the pseudo-first-order rate constant is given by

$$k_\Psi = \left[\frac{k_{ip} + k_{ion}K_d}{[M^+]} \right][X] \qquad (10\text{-}29)$$

The ion pair dissociation constant K_d can be determined by conductivity measurements independent of the kinetic measurements. Kinetic studies of k_Ψ as a function of either $[X]$ or of $[M^+]$ then allow the evaluation of both k_{ip} and k_{ion} by the use of either Eq. (10-28) or (10-29).

Experiments of this sort were carried out with $M^+ = Na^+$, and with $M^+ = Cs^+$ to obtain the following results:

$k_{ion} = 6.5 \times 10^4 \ M^{-1} \ sec^{-1}$ at 25°C for both Na^+ and Cs^+

$k_{ip} \ (Na^+, \ 25°C) = 80 \ M^{-1} \ sec^{-1}$

$k_{ip} \ (Na^+, \ -60°C) = 2.5 \times 10^2 \ M^{-1} \ sec^{-1}$

$k_{ip} \ (Cs^+, \ 25°C) = 21 \ M^{-1} \ sec^{-1}$

$k_{ip} \ (Cs^+, \ -60°C) = 1.0 \ M^{-1} \ sec^{-1}$

As required by the mechanism written, k_{ion} is independent of the identity of the cation. There are, however, several unusual features of the data for k_{ip}. First, it is highly unusual that the value of k_{ip} for the Na$^+$ studies is greater at -60 °C than at 25 °C. This implies a negative activation energy for the reaction. Second, the order of reactivity, R$^-$ > R$^-$Na$^+$ > R$^-$Cs$^+$ appears unusual in view of the expected greater ionic character of a C-Cs bond than a C-Na bond.

Both of these features are easily understood, however, if we examine the natures of the ion pairs more closely. From the data given in Table 6-6 we see that the Cs$^+$ ion pair is an intimate ion pair and probably remains as such at both temperatures studied. The Na$^+$ ion pair exists as both solvent-separated and intimate species, and the proportion of solvent-separated ion pair increases as the temperature is decreased. The $\Delta H°$ value for the conversion of [R$^-$Na$^+$] to [R$^-$//Na$^+$] is approximately -7 kcal/mole. Thus, the apparent negative activation energy results from the negative $\Delta H°$ for the conversion of intimate to solvent-separated ion pair in the case of the Na$^+$ reactions, and the higher reactivity of the Na$^+$ over the Cs$^+$ ion pair results from the fact that solvent-separated ion pairs are more reactive than intimate ion pairs in this reaction.

PROBLEMS

1. The rate constant for exchange of a methanol molecule hydrogen bonded to an amine with solvent methanol has been estimated to be 10^8 sec^{-1}. The second-order rate constant for the methoxide-ion-catalyzed H-T exchange of triphenylmethane in tritium-enriched methanol is $k_2 = 1.0$ x 10^{-9} M^{-1} sec^{-1}. From the data in Table 10-3, estimate the pK of triphenylmethane in methanol solution, assuming that the solvent exchange for this case has the same rate constant as that for the amine.

2. Assuming that k_1^D/k_1^T is independent of the structure of the hydrocarbon, calculate a^T from the observed isotope effect, $k_{obs}^D/k_{obs}^T = 1.52$, for the reaction of diphenylmethane in methanol at 100 °C. [See A. Streitwieser et al, J. Amer. Chem. Soc., 93, 5096 (1971).]

3. Are the results reported in Table 10-3 for 9-methylfluorene consistent with what you expect from the values of k_1, k_2, and k_3 for 9-methyl-2-dimethylcarboxamidofluorene reported on p. 254.

4. The isotope exchange, racemization, and isomerization reactions among the isomers:

(−)I—h   (+)I—h   (−)II—h   (+)II—h

(−)I—d   (+)I—d   (−)II—d   (+)II—d

have been studied in methanol solution, and are catalyzed by methoxide ion [D. J. Cram et al, J. Amer. Chem. Soc., 92, 4321 (1970)]. At equilibrium, II predominates over I by a factor of 7.5 at 25 °C.

When (−)I-h is 25% converted to II in CH_3OD, the reisolated I is found to be 64.1% exchanged and 67.3% racemized. The II formed is found to be 86.9% exchanged and 100% racemized.

When (−)I-d is 25% converted to II in CH_3OH, the reisolated I is found to be 23.2% exchanged and 21% racemized. The II formed is found to be 100% exchanged and 100% racemized.

a. Write a scheme similar to that on p. 251 for the various transformations and interpret the results in terms of rate constants in your scheme.
b. Suggest an explanation for the observed differences in the above two experiments.
c. Do these results provide any information about the validity of the assumption that k_2 is isotope independent in the reaction scheme of Eq. (10-10)?

REFERENCES

1. P. D. Bartlett and C. H. Stauffer, J. Amer. Chem. Soc., 57, 2580 (1935).

2. D. J. Cram, Fundamentals of Carbanion Chemistry, Academic Press, Inc., New York, New York, 1965, Chap. 4.

3. D. J. Cram, Fundamentals of Carbanion Chemistry, Academic Press, Inc., New York, New York, 1965, Chap. 3.

4. D. J. Cram, F. Willey, H. P. Fischer, H. M. Relles, and D. A. Scott, J. Amer. Chem. Soc., 88, 2759 (1966).

5. D. J. Cram, Fundamentals of Carbanion Chemistry, Academic Press, Inc., New York, New York, 1965, p. 102.

6. D. J. Cram, Fundamentals of Carbanion Chemistry, Academic Press, Inc., New York, New York, 1965, p. 88.

7. W. T. Ford, E. W. Graham, and D. J. Cram, J. Amer. Chem. Soc., 89, 689, 690, 4661 (1967).

8. C. D. Ritchie, J. Amer. Chem. Soc., 91, 6749 (1969).

9. C. D. Ritchie in Solvent-Solute Interactions, edited by J. F. Coetzee and C. D. Ritchie, Marcel Dekker, Inc., New York, New York, 1969, Chap. 4. (See also Vol. 2, Chap. 5.)

10. J. I. Brauman, D. F. McMillen, and Y. Kanazawa, J. Amer. Chem. Soc., 89, 1729 (1967).

11. M. Eigen, Angew. Chem., International Ed., 3, 1 (1964).

12. A. Streitwieser, W. B. Holleyhead, A. H. Pudjaatmaka, P. H. Owens, T. L. Kruger, P. A. Rubenstein, R. A. MacQuarrie, M. L. Brokaw, W. K. C. Chu, and H. M. Niemeyer, J. Amer. Chem. Soc., 93, 5088, 5096 (1971).

13. J. Smid and M. Szwarc, J. Amer. Chem. Soc., 87, 2764, 5548 (1967).

Supplementary Readings

A. Streitwieser and J. H. Hammons, Prog. Phys. Org. Chem., 3, 41 (1965).

M. Szwarc, Ions and Ion-Pairs in Organic Reactions, John Wiley and Sons, Inc., New York, New York, 1972, Chap. 1.

INDEX

Numbers in parentheses are reference numbers and indicate that an author's work is referred to although his name is not cited in the text. Underlined numbers give the page on which the complete reference is listed.

A

Abraham, M. H., 80(19), 93
Acetaldehyde hydrate, hydrolysis of, 192, 193
Acetate esters, hydrolysis of, 110-112
Acetic anhydride, pyridine catalyzed hydrolysis of, 6
trans-2-Acetoxycyclohexyl tosylate, solvolysis of, 83
Acid-base catalysis; see catalysis
Acid-base equilibria
    in aqueous solution, 161-164, 172
    standard states for, 162, 171
Acidity
    effect of polarizability on, 173
    gas phase, 173
    intrinsic, 173
    list of, 170, 172
    of hydrocarbons, 161-163, 169-173
    solvent dependence of, 164, 165, 171-173
Acidity functions
    comparison of various, 164, 165, 167, 168
    definitions of, 161, 164-166
    dependence on water activity, 168

[Acidity functions]
    for aqueous DMSO, 168, 169
    for carbonium ions, 161, 162, 165-168, 175, 176
    in acid-base catalysis, 187, 188
    in sulfuric acid, 167
    method of establishing, 165, 166
    use of, 164, 165
Activation, thermodynamics of, 39, 40, 43
Activation energy
    definition of, 38, 40
    negative, 264
Activity, of water in H_2SO_4, 167
Activity coefficients, 38-40, 43, 44-52
    in acid-base equilibrium, 162, 164, 165, 167, 168, 171
    in buffer systems, 183, 184
    interpretation of, 44
    of ionic solutes, 38, 44-48
    of nonelectrolytes, 44, 45, 47
    of solutes in DMF, 46-48
    of tetraalkylammonium ions, 45, 48
    of transition states, 39, 47, 48
    relative to DMSO standard state, 171
    role of, in acid-base catalysis, 186-191
    role of, in acidity functions, 162, 164, 165, 168, 171